Issues of Life & Death

Abortion
Birth Control
Capital Punishment
Euthanasia

Norman Anderson

InterVarsity Press
Downers Grove,
Illinois 60515

The London Lectures in Contemporary
Christianity, inaugurated in 1974, are designed
to stimulate Christian thinking on some of the burning
issues of the day.
The sponsoring body is the Langham Trust.
Any inquiries should be addressed to the Trustees,
2 All Souls Place, London, W1, England.

Second American printing, June 1978.

Published in the United States by
InterVarsity Press, Downers Grove, Illinois,
with permission from Hodder and Stoughton, Limited,
London, England.

InterVarsity Press is the book publishing division
of Inter-Varsity Christian Fellowship,
a student movement active on campus at hundreds
of universities, colleges and schools of nursing.
For information about local and regional activities,
write IVCF, 233 Langdon St., Madison, WI 53703.

ISBN 0-87784-721-5
Library of Congress Catalog Card Number: 77-74846

Printed in the United States of America

FOREWORD

WHEN I WAS invited to give the London Lectures in Contemporary Christianity for 1975, and asked to take "Issues of Life and Death" as my subject, I realised at once that it was somewhat foolhardy to accept, since the technical aspects of some of the subjects with which I should have to deal were right outside my competence. But I had been interested for a long time in the ethical problems inherent in questions such as abortion, euthanasia, the prolongation of life and the concept of a "Just Revolution", and had participated in the consideration of some of these subjects, and even the problems raised by genetic engineering, in one or another of the Interdisciplinary Discussion Groups which meet regularly at this Institute. All the same, my ignorance of science and medicine is so profound that an attempt to grasp the scientific details and background of much that I have discussed in these lectures involved venturing on to wholly foreign soil. Had it not been that I was so busy writing another book that I could not begin to delve into the scope of these lectures until it would have been unfair to withdraw, I might even have asked to be excused; and, had it been possible, I should certainly have preferred to postpone them for some months.

All the same, I must confess that I have found the subject fascinating. My method was the only one possible, I think, to one with my background: to turn to expert friends for advice about what I should read, to write draft lectures which some of these friends were kind enough to criticise, and to enlist help in dealing with technical points in the question period which followed each lecture. I can only say that they were characteristically generous in their response.

Among those who read one or more of the lectures and made most valuable suggestions, or helped me in some other personal way, I should like to record my special thanks to Dr Oliver

Barclay (Secretary of the Research Scientists' Christian Fellowship), Mr Gordon Barnes (Senior Lecturer in Zoology at Chelsea College), Professor R. J. (Sam) Berry (Professor of Genetics at the Royal Free Hospital School of Medicine), Dr John Vincent Edmunds (a Consultant Physician), Mr R. F. R. Gardner (a Consultant Gynaecologist), Dr Janet Goodall (a Consultant Paediatrician), Professor Donald Mackay (Professor of Communication in the University of Keele), Professor Eric Mascall (formerly professor of Historical Theology at King's College, London), Miss Elsie Sibthorpe (a Consultant Gynaecologist), Dr R. G. Twycross (of St Christopher's Hospice), Dr D. A. J. Tyrell (of the Clinical Research Centre, Northwick Park Hospital) and Professor Duncan Vere (Professor of Therapeutics at the London Hospital Medical College). I am also most grateful to Miss Shelagh Brown not only for typing and retyping these lectures but for a great deal of other help. My debt to the written work of Professor Gordon Dunstan (Professor of Moral and Social Theology at King's College, London) and to the authors of *Our Future Inheritance* (Professor W. F. Bodmer, of the Department of Genetics in the University of Oxford, and Dr Alun Jones, of *Nature*) will be obvious to all who read this book. But I must make it unmistakably clear that I alone am responsible for what I have written.

As usual, I have found difficulty in knowing how to refer to those authorities I have quoted. Scientists, I find, invariably refer to each other, in their books and papers, simply by surname, while other scholars follow different methods. So I have compromised by inserting their initials (together with Professor, Doctor, or whatever) on the first occasion when I have referred to them, and using their surnames alone in all subsequent references. For the rest, I have decided to let these lectures be published as I gave them, virtually *verbatim* (except that I usually had to omit part of what I had written in order to keep the lectures within any tolerable compass!), since I have not got the time to make the major additions which the production of a book on this subject *ab initio* would demand.

NORMAN ANDERSON

Institute of Advanced Legal Studies
University of London

1 The Sanctity of Human Life: Some Contemporary Problems

THE SUBJECT CHOSEN for this series of lectures — and I say "chosen" in the passive, because the choice was made partly by the committee which sponsors them and only partly by me myself — is "Issues of Life and Death". As such, they will range widely over a number of the most complex and controversial moral problems of today — problems concerned with the inception, the prolongation and the termination of human life. In my next lecture, for example, we shall be considering artificial insemination and genetic engineering — that is, the inception of life. After that, we shall turn to the deliberate prevention of the inception of life, whether by means of birth control, sterilisation or abortion — although abortion, of course, also represents the termination of life. And the last two lectures will be concerned exclusively with either the prolongation or the termination of life: whether life should always be prolonged to the maximum possible extent; in what circumstances (if any) it is right for it to be deliberately brought to an end by the hand of the person concerned or that of a doctor or someone else; and whether the violent termination of life by capital punishment, war or revolution — or, indeed, acts which put life in jeopardy in the cause of civil strife — can, in some circumstances, be justified.

All these questions are burning issues today, and a great deal has been said and written about each of them. Is there any point, then, in considering these subjects yet again, and is there anything new that I can say about them? I would reply that it seems most unlikely that I shall say anything whatever about any of these exceedingly controversial subjects which has not previously been said or written by someone else; but there may well be some value in considering them all together. Even this, of course, has already been done, in one way or another; but I do not know of any book written in the last few years which attempts to cover the subject in quite the same way, or from the same point of view, that I propose to follow. So I dare to hope that a fresh approach to this whole subject, in the severely summarised form that a series of five lectures necessarily imposes, may help some of us to clarify our minds and thinking on some very complex (and even agonising) moral problems which — directly or indirectly — concern us all.

But the question which immediately arises is why these problems are so complex and agonising. Most of us do not feel that the artificial insemination of animals — or even the possibility, whether now or at some future date, of submitting them to other forms of genetic engineering which do not involve cruelty — raises any acute moral problem, while we feel that to put an animal down when it is in distress or pain is positively praiseworthy. Again, most people see nothing wrong in the taking of animal life for food — however passionately we feel that this should be done in a way which reduces pain or fear to the absolute minimum. In other words, the criterion we apply to animals is not primarily the value of their lives as such, but the avoidance of any form of cruelty. This can, I think, be brought into sharper focus when we contrast it with the attitude adopted by Hindus to the cow, or by Jains, Buddhists and some Hindus to any form of animate life. To the Hindu the cow is sacred in itself, so it must never (in the commonly accepted view) be killed, even if it is wasting away from malnutrition or suffering from pain or disease which cannot be relieved; and to those Jains, Buddhists and Hindus who believe in a cycle of reincarnations, it is morally wrong to take the life of any living creature — including, in theory at least, a fish, reptile or insect — since by doing so one might be guilty of cutting short the present incarnation of a relation or friend. In all these cases it is clear that the criterion is not the physical suffering of the creature concerned, nor the reason for which its life might appropriately be terminated — whether to provide food, to prevent the spread of infection, to reduce over-population, or to benefit itself or humanity in some other way — but the intrinsic (or at least potential) value inherent in its life.

And this, of course, is precisely why it is issues of *human* life and death which raise acute moral problems. This is true, although in a different way and degree, both for the humanist[1] and the Christian. To the humanist, man represents the summit of evolution, unrivalled and unsurpassed in his potential; so his life is necessarily of unique value. To the Christian, on the other hand, it is God, not man, who is the lord of the universe, and the worth of man consists in the fact that he was created in the "image" and "likeness" of God and is capable of entering into a personal relationship with him; that in the incarnation God himself could, and did, partake of human nature; that for man physical death never, of itself, spells final extinction; that he can, even in this world, become a son of God; and that God has a glorious, eternal but largely unknown purpose for all those who, here on earth, have thus come to "partake in the divine nature". It is these facts, and their implications, which form the subject to which we must address ourselves.

First, then, let us attempt to summarise the value of human life as seen by the humanist who accepts, without qualification, a simplistic version of the latest theories about the evolution process. Astronomers now affirm that the universe contains something in the neighbourhood of a hundred million million million stars — with some thousand million extragalactic nebulae actually accessible to telescopic observation, every one of which may comprise some hundred thousand million planetary systems,[2] while previously unsuspected types of celestial

[1] I use this term in what I believe to be the generally accepted current sense: one who takes a high view of man but denies, or doubts, the existence of God.

[2] Cf. Fred Hoyle, *Frontiers of Astronomy* (William Heinemann, 1955), p. 83.

bodies, such as quasars and pulsars, are continually being discovered. Views about the universe, therefore, are likely to go on changing very considerably as our knowledge increases. The current theory, it seems, is that it began with "one big bang", and physics has no "explanation" for such a singular event.[3] But, however this may be, we are forcibly confronted with the relative insignificance of the planet on which we live and of the creatures that inhabit it. And this, in its turn, raises two further questions: whether there may not be a multitude of other sentient beings, similar or even superior to man, inhabiting a variety of other planets; and how one is to assess the value of a sentient being, relatively of less than microscopic size, in comparison with the unimaginable magnitude of the material universe.

But the second of these questions, at least, should not present any great problem to the humanist, for the issue involved is primarily a matter of the relative importance of quality and quantity. In Dr William Temple's vivid phrase: "I am greater than the stars, for I know that they are up there, and they do not know that I am down here." Viewed from another angle, it is like looking at a map and trying to assess the relative importance of vast expanses of arctic ice, desert sand and ocean waste in comparison with the insignificant dots which represent our major cities. It is to the Christian, rather than the humanist, that this question poses, at first sight, a much more difficult problem, for he inevitably wonders how the God who made the galaxies, and holds the universe, as it were, in the hollow of his hand, can "set his love upon the sons of men". And this wonderment is only partially answered when he begins to realise how infinitesimally small, unimaginably numerous and incredibly complex are the basic units of matter which make up the body of every human being.

But we must return to the question whether there may not be sentient beings, broadly similar to man, in many different parts of the universe. Scientists tell us that life (or, at any rate, life as we know it) — and here I quote Professor E. L. Mascall — "needs an atmosphere which contains plenty of oxygen but very little methane or ammonia", and that "certain

[3] Cf. J. Z. Young, *An Introduction to the Study of Man* (OUP, London, 1971), p. 363.

rather delicate and improbable relations need to be satisfied between the mass of a planet and its temperature throughout the course of its history if it is to provide a suitable theatre for biological evolution".[4] As a result, Sir Arthur Eddington, in 1928, would only admit that there might be "a few rival earths dotted here and there about the universe"[5] on which, conceivably, the evolutionary process might begin. It is in this context that Professor J. Z. Young writes: "The first question on this subject should perhaps be 'Did life originate at all or has it in some sense always existed?' This unfortunately is one of those great questions, like the origin of the universe, that we cannot handle yet." He then asks whether life arrived on our planet by migration from some other body; whether it was produced on earth by some "life force" operating outside the laws controlling matter as defined by physics; whether the laws that control the matter of the universe contain factors besides the known laws of physics, which dictated the necessity for life to begin (and presumably also to evolve); whether we can show that life may have arisen by the operation of forces known to operate in the terrestrial physical world; and whether life originated once only and on the earth, or once, or more often, elsewhere. In the present state of knowledge, he insists, none of these possibilities can be excluded. Nor, he concedes, can it be excluded that God or some vital force "so manipulated the materials of the early earth" that they acquired the property he terms "information storage" which enabled them to begin to evolve.[6]

Biologists will not necessarily, by any means, go all the way with a zoologist like Sir Julian Huxley or a palaeontologist like Teilhard de Chardin when they philosophise about the course of evolution, but it is interesting to note how these two men, starting from such very different standpoints, view the uniqueness of man. Huxley finds this in man's capacity for "conceptual thought", and then proceeds to explain, in detailed anatomical terms, why this "could have arisen only in . . . a terrestial offshoot of the higher Primates. Thus not

[4] Cf. E. L. Mascall, *The Importance of Being Human* (OUP, London, 1959), p. 2.

[5] *Nature of the Physical World*, p. 178.

[6] J. Z. Young, op. cit., pp. 367 f.

merely has conceptual thought been evolved only in man: it could not have been evolved except in man. There is but one path of unlimited progress through the evolutionary maze. The course of human evolution is as unique as its result . . . Conceptual thought on this planet is inevitably associated with a particular type of Primate body and Primate brain."[7] In much the same way, Teilhard de Chardin insists that "All the metamorphosis leading up to man reduces, from the organic standpoint, to the question of producing a better brain." But "if the being which issued in man had not been a biped, his hands would not have found themselves free to take over the prehensile functions from the jaws . . . It is thanks to the liberation of the hands through the adoption of biped movement that the brain has been able to grow."[8] Dr B. G. Campbell, moreover, sees the uniqueness of man not in what he terms "unconscious conceptualisation", in which some animals may be said to share, but in the fact that "human conceptual thought appears to be characterised particularly by its conscious nature", for in man alone, it seems, "thinking is a conscious process, and so, it follows, is conceptual thought" — aided enormously, of course, by articulate speech.[9] Somewhat similarly, Professor W. H. Thorpe sees the difference between the minds of animals and men as "one of degree — the degree of abstraction that can be achieved — rather than one of kind". It is when we come to "a recognition of abstract moral law, eternal values which are in themselves good," that, he feels, "we have reached a distinction which we can, for the time being at least, regard as fundamental".[10] All the same, Campbell virtually concludes his book *Human Evolution* with the words: "But what we have recounted is no more than a rough sketch of our subject. We know so little that we have been able only to glimpse a process of great duration and immense complexity. We might be forgiven for doubting that anything less than a miracle was needed to produce, by a process of adjustment from some strings of amino acids, a

[7] Cf. E. L. Mascall, op. cit., pp. 15 f.

[8] *Le Phénomène humain*, p. 187 — as translated by E. L. Mascall, op. cit., p. 7.

[9] *Human Evolution* (Heinemann, London, 1967), pp. 295 ff.

[10] *Biology, Psychology and Belief* (C.U.P., 1961), pp. 34 f. and 44 f.

Mozart or an Einstein."[11] But the fact that chiefly excites Huxley is that "the present situation represents a further highly remarkable point in the evolution of our planet — the critical point at which the evolutionary process, as now embodied in man, has for the first time become aware of itself, is studying the laws of its own unfolding, and has a dawning realisation of the possibilities of its future guidance or control."[12] This last point is one which we shall have to keep in mind, in a number of different contexts, in these lectures.

But to the Christian, as we have seen, the unique distinction and value of man must be expressed in very different terms. First, he believes — on the basis of divine revelation rather than scientific research — that it was God who, "in the beginning", created the universe *ex nihilo*, or "out of nothing"; for matter itself is not eternal. He also believes that it was God who subsequently caused the oceans and continents which make up this planet to bring forth "swarms of living creatures"; and then that it was he who, by whatever means, created man in his own "image" and "likeness", commanded him to "fill the earth and subdue it", and gave him a "dominion over the fish of the sea and over the birds of the air and over every living thing that moves upon the earth".[13] We are even told that God "crowned him with glory and honour" and "put all things in subjection under his feet".[14] But what does all this mean?

Let us begin with what is really the nub of the whole matter: the concept of man made "in the image" and "after the likeness" of God himself. A child who read these words would probably think first in terms of man's physical structure, and picture God in some sort of human form. Even among theologians, indeed, a few individuals have located the "image" of God, partly at least, in the fact that man stands erect! But the vast majority of theologians are "agreed that the *imago* is not to be found in any aspect of man's physical form"[15] — as is,

[11] Op. cit., p. 364.

[12] Cf. "The Evolutionary Process", in *Evolution as a Process* (eds. Huxley and others), p. 13.

[13] Cf. Hebrews 11:3; Genesis 1:1 and 20:8.

[14] Cf. Psalm 8:5 ff. and Hebrews 2:7.

[15] Cf. William Hordern in *Dictionary of Christian Theology* (ed. Alan Richardson, SCM Press, London, 1969), p. 202.

indeed, sufficiently apparent from the fact that we are specifically told that "God is spirit" and is invisible.[16] Now of course we often speak of him, almost inevitably, in anthropomorphic terms and metaphors; but *pure* spirit has not got a body even of the most ethereal sort. It is obvious, moreover, that man's physical frame is part of the natural creation: for man is depicted as made of "dust" like the rest of the animal kingdom, and as sharing with all animate life in its need for food and in the propagation of its species.[17]

Equally, however, we must not react against any such crude conception and try to isolate man's mind and spirit from his body. For the Bible "makes man a unity: acting, thinking and feeling with his whole being". So it is man as man, as Derek Kidner puts it, and not "some distillation from him", which is "an expression or transcription of the eternal, incorporeal creator in terms of temporal, bodily, creaturely existence".[18] It is in this context that Mascall insists: "Nowhere perhaps is the contrast between Christianity and most other religions more clearly seen than in its insistence on the fact that the body of a man is an integral part of him and not merely an accidental and temporary integument."[19] What is clear is that "in the opening chapters of Genesis man is portrayed as both *in* nature and also *over* it".[20] This is why he is depicted as having been given "dominion" over the rest of creation and having been told to "subdue" the earth.[21]

But how are we to define that aspect of man which is other than physical? The words we commonly use are either "soul" or "spirit" — and theologians sometimes argue as to whether human nature represents a trichotomy or a dichotomy. Those who take the first of these views regard man as made up of three parts, elements or aspects: body, soul and spirit; and they commonly quote, as biblical evidence for this view, 1 Thessalonians 5:23 ("may your spirit, soul and body be kept sound and blameless") and Hebrews 4:12 (where we are told

[16] John 4:24 and 1 Timothy 1:17.
[17] Cf. Genesis 2:7 and 19; 1:29 and 30; 1:22 and 28.
[18] Cf. *Genesis* (Tyndale Press, London, 1967), p. 51.
[19] Op. cit., p. 25.
[20] Derek Kidner, op. cit., p. 50.
[21] Genesis 1:26 and 28.

that the word of God pierces "to the division of soul and spirit, of joints and marrow" and discerns "the thoughts and intentions of the heart"). The first of these verses certainly seems, at first sight, to support a tripartite view of man; but the expression need not be understood in any technical sense, and may just be used as a comprehensive way of speaking of the entire man, in all aspects of his being; and in the second verse the distinction intended may well be between man's original Adamic nature, in which we all partake, and the new nature which the Christian receives in regeneration. Those who understand the nature of man in terms of a dichotomy, on the other hand, maintain that man is made up of only two parts or aspects: his physical body and that part of him which can be described as either soul or spirit. Many passages are cited by way of illustration, such as Ecclesiastes 12:7, Matthew 10:28, Romans 8:10, 1 Corinthians 5:3 and 5, 2 Corinthians 7:1, Ephesians 2:3, etc.; for all these verses appear to speak of man as made of both a physical body and an intangible element which can be described as either soul or spirit according to the context, the writer or the shade of meaning. This view seems to find at least partial support, moreover, in the fact that death is sometimes described as the giving up of the "soul"[22] and sometimes the giving up of the "spirit"[23]; and it is significant that the Greek word used for "souls" is simply translated "lives" in Acts 15:26 and that the word translated "spirit" need mean no more than breath or "life principle". It must also be remembered that in the Old Testament the Hebrew words *nephesh* and *ruach* both have a number of different meanings and are used of animals as well as men;[24] and in the New Testament, too, the Greek words *psyche* and *pneuma* have different meanings in different contexts. It can, I think, safely be said that when the adjectival form of the first word (*psychikos*) is contrasted with that of the second (*pneumatikos*) in reference to man, the former refers to man as he is by nature (or, in 1 Corinthians 15:44, in his physical body), while the latter is used of man when indwelt by the Spirit of God (or, in 1 Corinthians 15:44, in his resurrection body). But it seems to me that a very strong

[22] Cf. Genesis 35:18; 1 Kings 17:21.
[23] Cf. Psalm 31:5; Luke 23:46; Acts 7:59.
[24] Genesis 1:20, 21 and 24 and Genesis 6:17, respectively.

case can be made for insisting that man is, in fact, a psychosomatic *unity*, in which the physical and non-physical are complementary aspects rather than separable elements or parts. This is why the day of resurrection is viewed in terms of the dead being raised to life not as disembodied spirits but in what St Paul terms "spiritual bodies".[25]

How, then, are we to understand the statement that God made man "in his image" and "after his likeness"? Not infrequently, a sharp distinction has been made between the two terms. St Augustine, for example, seems to have held that the former refers to man's intellectual and the latter to his moral qualities. Many Roman Catholic theologians, again, regard the word "image" as referring to man's "natural" gifts — that is, the fact that he is a rational and morally responsible creature — and "likeness" as a reference to gifts with which man was, at the creation, "supernaturally" endowed — that is, his basic righteousness or innocence before he fell into sin. But it is, I think, most improbable that the two phrases were in fact intended to convey distinct and different meanings, and much more likely that they should be taken as a composite expression; and this appears to be confirmed by the fact that the term "image" is used alone in Genesis 9:6, and the term "likeness" in Genesis 5:1 and James 3:9, but without any discernible difference of meaning.

All the same, it seems clear that this composite phrase has both a wider and a more restricted meaning. In its wider sense the fact that man alone in the physical world was made "in the image" and "after the likeness" of God must surely mean that man is, in an unique degree, a rational being, capable of conscious, conceptual thought, which he can communicate to others by articulate speech; that he is a morally responsible being, with an abstract sense of right and wrong; that, while himself a creature who is essentially dependent on his creator, he has been given a freedom of moral choice, a delegated authority and an ability to enter into, and take his part in, the creative activity of God; and, above all, that he is a creature with whom God can communicate and who can enjoy a

[25] For this whole subject, cf. B. F. C. Atkinson, *Life and Immorality* (The Phoenix Press, Taunton), where there is a detailed analysis of the Hebrew and Greek words for "soul" and "spirit".

personal relationship with his Creator. In the narrower meaning of the phrase, on the other hand, it seems equally clear from Ephesians 4:24 and Colossians 3:10 that man was originally created in the "likeness" and the "image" of God in the sense of a basic righteousness and holiness and an innate knowledge of his Maker. Then came what theologians call the Fall, in which man, by deliberate and wilful disobedience, lost God's image and likeness in the narrower meaning of the phrase, for he is no longer basically righteous or holy, nor does he enjoy a true knowledge of God. Yet in the wider meaning of the phrase it seems clear that, while God's image and likeness in man has been marred, indeed, it has not been wholly lost. As a rational creature man is today often in error, but he still has the power of conceptual thought. As a moral being he frequently does what is wrong, he never attains perfection, and his very conscience may become "seared as with an iron"; but the requirements of God's law are still, basically, "written on his heart", and he knows something of God's judgment on sin.[26] He still knows the meaning of love; he still, within limits, has a freedom of choice; and he still enjoys a delegated authority over nature. After all, it was after the Fall that the prohibition of shedding man's blood, precisely because God had "made him in his own image", was enunciated,[27] and that the evil of cursing men, "who are made in the likeness of God" was denounced.[28] What is more, God can and does still communicate with man, and man can and does still become a son of God by regeneration and adoption.[29] When this takes place, moreover, God begins to restore in him his own image and likeness in the narrower sense of that phrase,[30] by a progressive renewal of the holiness, righteousness and true knowledge of God that man lost in the Fall. But it is only in Christ himself that the "express image" of God's person, or the "very stamp of his nature",[31] can be seen, for he alone was (and is) both truly God and truly man. So it is only through him that fallen men and women can become "partakers in the divine nature";[32]

[26] Cf. Romans 2:15 and 1:32.
[27] Cf. Genesis 9:6.
[28] Cf. James 3:9.
[29] Cf. John 1:12 f. and Galatians 4:5.
[30] 2 Corinthians 3:18.
[31] Hebrews 1:3. [32] 2 Peter 1:4.

and it will be only when they see him as he is that, at last, they will become really like him,[33] and he will be "the first-born among many brethren".[34] There is a very real sense, therefore, in which it is only in Christ that we can see humanity as God intended it to be, and only in him that we can become truly human.

To what extent, then, does the biological view of man, as seen through the eyes of the contemporary scientist, run parallel to, or part company with, the theological view derived from the Bible? I must not digress, in this context, into a discussion of the question whether the statement in Genesis 2:7 that God "formed man of dust from the ground, and breathed into his nostrils the breath of life; and man became a living being" should be understood in terms of a divinely-guided evolution of man's physical frame from common chemicals ("the dust of the ground"), through all the mutations the evolutionist suggests, and then a new creative act by which God constituted Adam as the first man in the full sense of that term; or whether it should be taken to mean a direct creation of man as a psycho-somatic organism (body and soul, flesh and spirit) from these same basic chemicals — although incorporating, no doubt, most of the anatomical features which had already been developed in a succession of other mammals. If such direct creation is, in fact, what we should believe, then I cannot myself see how science could conceivably disprove it; for none of the links in the evolutionary chain, whether apparent or real, could possibly demonstrate the precise way in which God created the human race as such. But many Christians who wholeheartedly accept the biblical revelation take the other view, and either accept the phrase "dust from the ground" as itself denoting one of the higher anthropoids or as referring to the beginning of the whole evolutionary process.

But here a further note of caution must be registered. To return to the evolutionary thesis pure and simple, Young states that most biologists probably agree:

> that the question of which fossils are truly those of a man cannot ever be settled. If evolutionary change is gradual it is

[33] 1 John 3:2.
[34] Romans 8:29.

> not simply difficult to solve such questions, it is *impossible* in principle. Apart from the difficulty of the gradualness of change there is the question of definition, which is bound to be to some extent arbitrary. A sudden beginning for man might be imagined as the result of the operation of some creative process, not found elsewhere. The endowment of the race by some outside force, agent or God with a new entity "the soul"[35] would be such a process, and of course it is not to be ruled out arbitrarily that this is what happened. Again there might, conceivably, have been a sudden genetic mutation, in one person, leading his descendants to be not only highly successful but also fertile only among themselves. We could then say he was the first man. A possible suggestion of this sort is that the size of the brain was suddenly doubled by the addition of one more cell division before the end of the multiplication of neurons.[36]

I presume, moreover, that a "sudden genetic mutation, in one person" might not only start a new race which was "highly successful", but one between which and its predecessors there were certain fundamental differences. But Young himself, in company with most biologists, believes that *Homo sapiens* gradually evolved.

> *Ramapithecus* was more like a man than were the Dryopithecines, *Australopithecus* was more like a man than *Ramapithecus*, and so on. The series is very incomplete and the existence of sudden small jumps cannot be ruled out, but the evidence is compatible with gradual change . . . Perhaps it is at the last stage of all that we should be most prepared to expect the occurrence of some sudden change. Man is indeed very different from all other animals, even from such fossils as those of *Australopithecus*. We feel that we must be in some way widely divergent from them, the product of some special and perhaps sudden process. This feeling is

[35] Though this is not really the appropriate term; for the word "soul", as we have seen, is used in Genesis 1 of animals as well as man.

[36] Op. cit., pp. 456 f.

egocentric but cannot be dismissed as wholly irrational. Modern man *is* very different.[37]

Commonly, however, biologists trace his origin through *Homo erectus* (the Java and Pekin men), *Homo sapiens neanderthalensis*, to *Homo sapiens sapiens*.[38] The successive stages are, moreover, normally assessed in both anatomical and cultural terms: the growth in the size and structure of their brains, and the development in what they could make and do. So it is obvious that a major problem remains as to where we should place what we may term Adamic man; for to the Christian it is neither the size of the brain nor the ability to talk or make tools which decides whether a creature is truly "human", but whether he has the capacity to enter into a personal relationship with God. In any case, the all-important fact, in the context of this lecture, is the creative act of God — whatever precise form it may have taken — by which our first parents were made truly human, with all that this involved before the Fall.

But what likelihood is there, on the purely evolutionary thesis, that man himself will go on evolving into something very different — into some sort of superhuman creature quite distinct from the men and women we know today? It is true that F. Le Gros Clark, after his statement that the "human" brain has "not appreciably changed in size for about 200,000 years", immediately adds that "there seems to be no evidence that man's brain is undergoing any further evolutionary expansion — or that it is even likely to do so".[39] Yet it is in this context that Mascall justly remarks that "even if the process of natural evolution has come to a virtual stop with the arrival of a rational species, the very arrival of that species opens up the prospect of a vastly accelerated... development of an entirely new kind, that is to say, a development which will be consciously and deliberately planned by the rational species itself."[40]

Now it will be our task in later lectures in this series to explore the degree to which it is already possible, or may become

[37] Loc. cit., p. 457. [38] Loc. cit., p. 458.

[39] Cf. "The Structure of the Brain in the Process of Thinking", in Laslett, ed., *The Physical Basis of Mind*, p. 23.

[40] Op. cit., p. 21.

possible, for man to modify and even manipulate his own genetic material — and, indeed, to discuss the ethical problems this may involve and how far one possibility or another may, or may not, be morally right. Few, for example, would deny that it would be perfectly permissible to improve our genetic heritage — physical, mental and moral — by the elimination of dangerous genes, provided this can be done without taking unacceptable risks and in a way which does not involve any overriding moral objections; for why, otherwise, should we not extend the protection which we always try to accord individuals, or the race as a whole, against noxious germs to a similar protection against noxious genes? Even if this proves to be possible and seems to be legitimate, however, have we any reason to think that the human species will radically change in its essential nature? To the humanist, I suppose, the only obstacle is to be found in the dubious likelihood of any further expansion in the cerebral surface of man's brain. But for the Christian a convincing answer can, I think, be found in the Incarnation, which provides us, in Mascall's words, with solid reasons for holding that, whatever further developments the race may undergo, "man is meant to remain man; for if God has himself become man . . . he has sealed human nature with a certificate of value whose validity cannot be disputed".[41] The Incarnation was not an episode in which he who "though he was in the form of God, did not count equality with God a thing to be grasped, but emptied himself (or 'made himself nothing') taking the form of a servant, being born in the likeness of men",[42] only to return to his original state after his passion and exaltation. On the contrary, the Christian doctrine is that he who was always God became truly man without ceasing to be God; and that he is now ascended "to the right hand of God" — and is, indeed, "in the midst of the throne" — without ceasing to be man.

It is in this context that Mascall cites passages from *The Silence of St Thomas*, by Josef Pieper, to demonstrate the direct link which exists between the very idea that things have "natures" and the notion that they are "the products of a

[41] Op. cit., pp. 22 f.
[42] Cf. Philippians 2:6 f.

form-giving thought".[43] It is true that the suggestion that "there could be no such 'nature' unless it were creatively thought" is strangely alien to modern Rationalism; but Pieper shows that Jean-Paul Sartre, at least, is perfectly conscious of "a sad lack of clear and logical thinking, when the concept of creation is abandoned but not the habit of talking about the 'nature of things', as though on that point nothing had changed". It is precisely because Sartre denies the existence of any creative intelligence which could have designed man and all natural things that he also denies that there is any "nature" in things which are not manufactured and artificial. "There is no such thing as human nature," he insists, "because there exists no God to think it creatively." And Pieper pertinently concludes that "It is superficial, unreasonable, and even absurd to maintain that there is a 'nature' of things, anterior to existence, unless one holds at the same time that things are created."[44]

Up to this point, it may perhaps have been noticed, I have refrained from any mention of "immortality" as one of the characteristics of man made in the image and likeness of God. This omission was both conscious and deliberate, since the whole subject of immortality, and what in fact it means and implies, seem to me to demand separate consideration. To begin with, I am not myself fully convinced that the concept of the *essential* immortality of the human soul is not derived primarily from Greek philosophy rather than biblical teaching. It is distinctly arguable, I should have thought, that the first phrase in the New testament description of the "only Sovereign, the King of kings and Lord of lords" as the one "who *alone* has immortality and dwells in unapproachable light"[45] should be taken as literally as the second. In this case the word translated "immortality" is literally "deathlessness"; but in the closely parallel doxology earlier in the same letter ("To the King of ages, immortal, invisible, the only God, be honour and glory for ever and ever")[46] the word for immortal is

[43] Op cit., p. 23 — using the word "nature" in the sense of "the properties essentially pertaining to anything and giving it its fundamental character". (S.O.E.D.).

[44] Cf. pp. 51–3 and 94 f.

[45] 1 Timothy 6:16.

[46] 1 Timothy 1:17.

literally "imperishable" — and it is interesting to note that these two Greek words are both used in 1 Corinthians 15:53 when the Apostle, writing about the resurrection of Christians, insists: "For this perishable nature must put on the imperishable, and this mortal nature must put on immortality. When the perishable puts on the imperishable, and the mortal puts on immortality, then shall come to pass the saying that is written: 'Death is swallowed up in victory'." It is no surprise, therefore, when we read in 2 Timothy 1:10 that it was "our Saviour Christ Jesus who abolished death and brought life and immortality [here literally 'imperishability'] to light by the Gospel"; and Dr W. Hendriksen aptly comments: "For the believer immortality is therefore a redemptive concept. It is *everlasting salvation.* For God it is *eternal blessedness.* For while the believer *has received* immortality, as one receives a drink of water from a fountain, God *has* it. It belongs to his very being. He *is* himself the Fountain".[47]

But what can we say about man as man, from the very beginning? We are told that God "commanded the man, saying 'You may freely eat of every tree of the garden; but of the tree of the knowledge of good and evil you shall not eat, for in the day that you eat of it you shall die'." And when Adam disobeyed this explicit command, God's judgment was expressed in the words: "In the sweat of your face you shall eat bread till you return to the ground, for out of it you were taken; you are dust, and to dust you shall return."[48] Almost immediately, moreover, we read: "Then the Lord God said, 'Behold, the man has become like one of us, knowing good and evil; and now, lest he put forth his hand and take also of the tree of life, and eat, and live for ever' — therefore the Lord God sent him forth from the Garden of Eden, to till the ground from which he was taken."[49]

These are mysterious words, and he would be a bold — indeed a foolhardy — man who ventured to dogmatise about their meaning. But the "tree of life" reappears in Revelation 22:2, and no commentator takes the expression literally there. It suffices for our purpose, however, to emphasise that the

[47] *The Epistles to Timothy and Titus* (Banner of Truth Trust, London, 1959), p. 208.

[48] Genesis 2:16 f. and 3:19.

[49] Genesis 3:22 f.

basic point is that our first parents committed an act of wilful disobedience which broke their communion with God, brought their state of innocence to an end, replaced it with a consciousness of sin and guilt, and eventually involved physical death as we now know it. St Paul's commentary on this story of the Fall first insists that "sin came into the world through one man and death through sin, and so death spread to all men because all men sinned"; then depicts the resultant reign of sin and death; and finally declares that "If, because of one man's trespass, death reigned through that one man, *much more* will those who receive the abundance of grace and the free gift of righteousness [or 'justification]' reign in life through the one man, Jesus Christ."[50]

But what confuses the issue so often, to our limited understanding, is that, when we read about "death" in the Bible, one or both of two different — although closely related — concepts may be involved: physical death and that spiritual death which is separation from God in whom, alone, is the fountain of life. It is clear that physical death, as we now know it, is not regarded in the Bible as just an inevitable concomitant of man's physical constitution, but as the judgment of God on his sin. In Dr Martyn Lloyd-Jones' words, that "does not mean that Adam by creation was in a condition which was already immortal, and that he would continue as such to all eternity. He was perfect, but he was not glorified. Adam had still to achieve immortality, but there was no principle of death in him."[51] In point of fact the Bible only speaks of two men — Enoch and Elijah — who did not experience physical death. Both of them, of course, knew the spiritual alienation or death of sin, and both of them experienced the "free gift" of life which was later to be made manifest in Christ. In the case of Enoch, indeed, the record states, with sublime simplicity, that "Enoch walked with God; and he was not, for God took him".[52] And although the hope of immortality is far from

[50] Romans 5:12 and 17.

[51] *Exposition of Romans Chapter 5* (Banner of Truth Trust, London, 1971), p. 195. In other words, immortality is not an inherent quality of the human soul. Instead, the resurrection of the dead represents an act of divine power; and immortality, in the full sense, is enjoyed only by those who participate in the divine nature.

[52] Genesis 5:24.

prominent in the Old Testament, it is interesting to note that the Hebrew word for God "taking" Enoch reappears in Psalm 49:15 and Psalm 73:24. In the first the Psalmist writes: "God will ransom my soul from the power of Sheol (the abode of the dead), for he will *receive* me"; and in the second "Thou dost guide me with thy counsel, and afterward thou wilt *receive* me to glory".[53] But it is only in the New Testament that we find explicit teaching about that "eternal life" (the very definition of which is to *know* "the only true God and Jesus Christ whom he has sent")[54] that the Lord Jesus died to give us.

But what of those who refuse this gift and reject the Saviour? About their eternal destiny we can speak only in terms of what the Bible tells us. There is no statement that I know which asserts that there is a part of man which is essentially "immortal", and which continues to exist for ever as a disembodied spirit. On the contrary, St Paul insists that in the resurrection the redeemed will be given a new "spiritual" body, through which they will be able to express their true personalities in the life to come much more perfectly than has been possible through their physical bodies here on earth.[55] In this physical body, he says, we "groan" and "sigh with anxiety"; yet what we long for is not to be "unclothed" or "found naked", but to be "clothed upon" with our spiritual bodies, or our "heavenly dwelling", so that "what is mortal may be swallowed up by life".[56] But, again, what of those who have persisted in refusing God's gift? That a day is coming when all men will be "raised" from the dead, and when God will judge the world with perfect justice and complete understanding, is clearly taught; and so, it seems to me (agonising though this should be to every one of us) is the final, irreversible judgment on those who stand condemned. What must, I think, be regarded as the traditional view sees this condemnation in terms of an eternal conscious separation from God; but it would, I believe, be equally true to biblical teaching to express it in terms of a death and destruction which is "eternal" in the sense that it is final and irreversible.[57]

I am afraid that I have digressed at a number of points from

[53] Cf. Kidner, op. cit., p. 81.
[54] John 17:3.
[55] Cf. 1 Corinthians 15:35–57.
[56] Cf. 2 Corinthians 5:1–5.
[57] Cf., again, B. F. C. Atkinson, op. cit

the main tenor of my argument — although the digressions have seemed to me to have some relevance. But at this point it is essential to draw the whole subject together and set the stage, as it were, for the lectures which are to follow. So I hope that I have at least left you in no doubt that our basic concept, to which we shall continually have to return, is the unique significance and value of the human personality: man made in the image and after the likeness of God himself; man as he was created, as he now is, and as God would have him to be.

It is in this light that we must approach the problems of artificial insemination, genetic engineering, birth control, sterilisation, abortion, and all that the next two lectures are to cover; and also, of course, the prolongation of life, transplant surgery, euthanasia, suicide, capital punishment and the termination of life in the course of violence, revolution or war. Some of these subjects could justly be regarded by the humanist as man taking a conscious and active part in his own evolutionary development — which is, of course, the reason why I have given some attention to this topic in this lecture; and they may certainly, I think, be considered by the Christian in the context of man attempting to obey the divine command to "subdue the earth", to co-operate with God in his on-going creative purposes, and to "replenish the earth" in an intelligent way. But whereas, in the animal kingdom, there would appear to be the widest scope for experimentation — provided always that this does not involve unnecessary pain, unacceptable exploitation or wrongful neglect of the principles of conservation — the situation is very different in the case of man. With an animal, for example, artificial insemination seems to involve no moral problems, but in the case of human beings a number of moral, aesthetic, psychological, social and even legal considerations must, of necessity, be taken into account; and this is also true of genetic engineering, sterilisation, euthanasia and much else. In other words, we must grapple with the problem of how far the command to "subdue the earth" includes man's own genetic and psychological make-up. And in this, and a number of other respects, the approach of the humanist may differ widely from that of the biblically minded Christian.

So we come back to where we began: the essential nature of man. On the one side of his nature man is indubitably part of the animal kingdom; but, in a unique degree, he is a rational creature, capable of conceptual thought. Partly for this very reason, he has the God-given ability to transform — or at least control — nature, and to participate in God's creative purposes. More important, from the spiritual standpoint, he is a moral creature, who can in part distinguish, even in his fallen condition, the difference between right and wrong. And still more important, he is a creature with whom God himself can communicate. As Professor G. R. Dunstan puts it, God is continually the Creator, "the source and giver of life. He creates man 'in his own image'; that is, with a potential for an identity capable of awareness of God, capable of a freely-willed response to the awakening or call of God, and therefore capable of a consequent likeness with God, of being stamped with his mark, his 'character', his image."[58] It is in this context that Mascall asks why it is that we need to hold that in man there is a distinct spiritual component which is not found in any sub-human creature, and replies:

> If we admit that God is pure Spirit and at the same time recognise that man is not a *pure* spirit but has a body which is continuous with the rest of the material creation, have we any real alternative, believing as we do that man is made in the image of God, than to hold that the way in which God made man was by uniting a physical organism — which as a physical organism did not differ in kind from other physical organisms — with a created spirit which, without suppressing the animal and vegetal functions of the physical organism, could subsume them into and make them subservient to its own supra-physical life?

This, he insists, is not the view

> that man is in his essence a pure spirit whose habitation of a material body is a temporary and, on the whole, a tiresome episode ... But neither is it, on the other hand, the view which would simply identify man with his body ... It is the view that man is a unique and highly complicated being composed of a body which is more elaborate than, though

[58] *The Artifice of Ethics* (SCM Press, London, 1974), p. 70.

not necessarily different from, that of any other of the primates, and a soul which, although it is in itself a purely spiritual entity, is not the kind of spirit that can function fully and freely on its own, since it is made for the express purpose of animating a material body and in fact of animating that particular human body with which it is united.[59]

So, as we approach the exceedingly complex and difficult questions to which we must turn in my next lecture, the Christian must insist that, while man is an integral part of the animal creation, he is also essentially distinct from it; that to speak of "human nature" as in a category of its own is both meaningful and true; that, however much it may be right — and even mandatory — to attempt to improve the health, intelligence and enlightenment of our race, human nature as such cannot be radically changed; and that any act or process designed to "dehumanise" man stands, *ipso facto*, condemned. Perhaps, then, we can most appropriately end this lecture with words from the Old Testament, as quoted — and then expanded — in the letter to the Hebrews:

> "What is man, that thou rememberest him, or the son of man, that thou hast regard to him? Thou didst make him for a little while lower than the angels; thou didst crown him with glory and honour; thou didst put all things in subjection beneath his feet." For in subjecting all things to him, he [God] left nothing that is not subject. But in fact we do not yet see all things in subjection to man. In Jesus, however, we do see one who for a short while was made lower than the angels, crowned now with glory and honour because he suffered death

—and suffered death, we are told, in the course of bringing many sons to glory".[60]

[59] Op. cit., pp. 27 and 28.
[60] Hebrews 2:6–10 (NEB).

ADDENDUM

The condition of the dead between death and the day of resurrection is a question on which opinion is sharply divided. The most commonly held view, I think, is that their disembodied spirits are alive and conscious — and "with Christ", of course, in the case of believers. Frankly, I am not happy about this view, and I am by no means convinced that the parable of Dives and Lazarus, the account of the Transfiguration, or the story of the witch of Endor, compel us to accept it. Another view is that believers who have died are "alive" indeed (cf. Luke 20:37 and 38) but "asleep in Jesus" until they receive their new spiritual bodies, without which they cannot experience *conscious* life. And what may perhaps be regarded as a variant of this view, suggested by C. S. Lewis,[61] takes the form of a simile from inanimate life which runs something like this. A composer conceives a concerto with fully orchestral score, but initially commits it to paper only in terms of a piece of piano music. This is played and becomes well known, but the paper on which he wrote it is not duplicated and is eventually destroyed. Then, after an interval, the composer writes out the full score which he always had in mind. This is far more wonderful than the original, but unmistakably the same piece. What, then, could be said about the concerto in the interval between the notation for the piano and its transcription in the full orchestral score — except that it was certainly "alive", but not embodied? Finally, there is a third view, to which I myself incline. When we die we pass out of a space-time continuum into a realm where time is merged in eternity; so might it not be true that those who die in Christ are immediately with him, in their resurrection bodies, at the Advent — which, while still future to those of us who still live in time, is to them already a present reality?

[61] Cf. *Transposition and other Addresses* (Geoffrey Bles, London, 1949), pp. 13 f. and 20.

2 Genetic Engineering and Artificial Insemination

I DEVOTED ALMOST all my time in my last lecture to a consideration of the nature of man and the essential value of human life, since this seems to me to constitute the indispensable background to any discussion of the "issues of life and death" which will concern us in this and in my succeeding lectures. To the humanist, we saw, man may be regarded as the apex of the evolutionary process, while to the Christian the fundamental fact is that man was created, by whatever means, "in the image" and "after the likeness" of God. To the humanist, again, the exciting point has been reached at which man can consciously take a part in furthering his own evolution, while to the Christian some of the undreamed of possibilities — actual or potential — which are being explored today by medical research may be regarded as an opportunity for man to fulfil, in a new and more intelligent way, the divine command to "fill the earth and subdue it". But those engaged in such medical research are, for the most part, profoundly conscious of the fact that their work must be governed by ethical as well as by purely scientific considerations. If the use of human beings to test out the efficacy — and the possibly dangerous side-effects — of new drugs is regarded by the medical profession as irresponsible unless it is done in accordance with strictly defined criteria, then how much more must even the most *avant garde* research worker quail before the terrifying possibilities inherent in any tinkering with the genetic future of the human race or of its individual members. And these possibilities will, of course, be at least equally relevant to the Christian — together, perhaps, with certain other considerations which may well weigh more heavily with him than with the ethical humanist, or even be unique to the Christian as such.

So, before I turn specifically to the subject of my present lecture, let me lead in to the problems we have to face in all my remaining lectures by two quotations. The first comes from a Review Article on "What is Human Life?" by Mr C. G. Scorer, a Consultant Surgeon:

> Man's confidence in the possibilities of his own technological achievements is almost unlimited. There is nothing in this world he cannot do if he really sets himself to do it. As he literally reaches for the stars, as he increases his control over his environment, so he sets about securing his own destiny by remodelling himself. Medical progress has recently pushed open a new door. We have become fascinated by the possibilities of what we may soon be able to do with our own species. Curiosity beckons us. Politicians envisage new possibilities of power. Surely along this way lies the solution to some of the world's most pressing problems? But can we be certain about this — certain that we are not raising even greater problems than we are trying to solve? If God is Creator and Lord of this world — and of Man in particular — are there limits which need to be set to man's dominion?[1]

And the second comes from a monograph entitled "Is life really sacred?" by Dr Paul W. Brand, where he reminds his fellow doctors that

> we as physicians are facing a new problem. It is the dilemma that is forced upon us by the fact that today we sometimes have more power over life and death than we would choose to have. We have the power to prolong life when life seems to have lost its meaning, and we have the power to terminate

[1] *In the Service of Medicine*, April 1971, p. 21.

life without suffering. We wonder whether we have the right to choose life and death for our patients. If we are not to choose, then whose judgment do we accept?[2]

It is precisely in regard to points such as these that the ethical problems posed by the rapidly expanding horizons of scientific and medical knowledge press most heavily upon us in the context of the subject matter of these lectures. What new obligations are laid upon those most intimately concerned by the prospect that human behaviour, and the very constitution of human beings, may be shaped by scientific means? To quote Professor D. M. Mackay: "We want to do good, if any good may be done with our new knowledge; yet we are unhappily aware of the ethical risks in the manipulation of human beings, even with the best of intentions. Torn by the claims of compassion and of respect for the individual, men of goodwill look to one another for guidance, and find it hard to see a clear way forward."[3]

What is clear, I think, is that this way forward must lie somewhere between the Scylla and Charybdis of two opposite dangers. On the one side we have voices which continually warn us that almost any form of genetic engineering is a menace to humanity, is contrary to nature, and involves ethical and spiritual problems beyond our competence to solve. Let us, then, be content with doing our best to care for the casualties of nature as we know it, rather than make any rash attempt to manipulate nature itself. And on the other side there are the voices which must sound so seductive in the scientist's ear urging him ever onward — in terms of compassion for suffering humanity, the fascination of new discovery and even the allurement of personal fame.

Nor is the Christian exempt from any of these dangers. He should, indeed, be more conscious than the humanist that man is both finite and sinful; and this should serve as a continual reminder of the catastrophic mistakes he can make, even with the most laudable intentions, and a perpetual warning against

[2] Published in 1973 under the auspices of the Christian Medical Fellowship (56 Kingsway, London WC2).

[3] From a paper on "Biblical Perspectives on Human Engineering" written for a Conference on "Human Engineering and the Future of Man" (Wheaton College, Illinois, July 1975).

that arrogance of spirit which forgets the obedience and dependence which the creature always owes to his Creator. But this does not give him any excuse for holding back when new possibilities of doing good to his fellow men or to the human race are opening up before him. To do this would be to be guilty of disobeying the divine command to "subdue the earth", of laying up his "pound" in a napkin instead of using it to God's glory, and of failing to do the good that he knows. But are there no principles by which he can steer his way over these uncharted seas? First and foremost, I think, we must accept the fact that "nature", as we know it, can provide us with no adequate criterion, for "the whole creation has been groaning in travail together until now"; and although the created universe will not be "set free from its bondage to decay" until the final "revealing of the sons of God" — for which, the Apostle tells us, it is waiting with "eager longing"[4] — yet it is surely our duty to do all we legitimately can for our fellow-men, and the very world in which we live, in the meantime. Our God is not a pagan deity who jealously guards the secrets of nature, but the "Father of our Lord Jesus Christ" who graciously (and progressively) reveals them to man made in his image — partly in order that he may exercise the dominion with which he was entrusted. We must always remember, however, that this is a delegated authority, which can only rightly be exercised according to the Creator's revealed will. This means, Mackay suggests, that Christians should be champions of science, but *critical* champions, who should oppose research only if it infringes Biblical principles or would prevent them doing something better, something more glorifying to God. For it is his glory which should always be our supreme concern — in terms of the highest good of individual men and women, of that family structure which may be said to represent a "creation ordinance" or part of our "Maker's directions" for his creatures, and of human society as a whole.

In this second lecture the course I propose to adopt is broadly as follows. First I shall set out, in as objective (and, indeed, as severely clinical) a way as I can, the relevant facts and possibilities — whether present, imminent or remote — in

[4] Cf. Romans 8:19 .

regard to each of the points we have to consider. Next, I shall attempt to weigh the actual or potential benefits against the possible objections or dangers involved. Finally, I shall try to outline the ethical considerations which, as it seems to me, must inform the Christian conscience in assessing the subject — and even to hazard, in some cases, a judgment of my own.

Now it is obvious that an attempt by man to take a hand in his own evolutionary process — or rather, let us say, to co-operate in a new and more intelligent way with the creator in his creative purposes for the human race — can adopt either a positive or a negative approach. On the negative side attempts may be made to minimise the birth of children who will themselves suffer from serious physical, mental or moral defects or who are likely to transmit such defects to their progeny; or, again, to prevent the birth of more children than the natural resources of the earth can sustain, or than their parents can properly provide and care for. It is with this approach that we shall be concerned next week, when our subject is to be birth control, sterilisation and abortion. Equally, however, it is possible to adopt a much more positive approach and to concentrate on the steps that can (or might) be taken to ensure that healthy children are in fact born to those who can care for them, and that genetic defects are eliminated, or at least controlled, to the greatest possible extent. It is this approach that we have to consider in some detail this evening under the title of "Genetic Engineering and Artificial Insemination".

The very term "Genetic Engineering", understandably enough, fills many people with intense alarm. This is partly, no doubt, because of the mental picture to which journalistic references to "test-tube babies" — or, indeed, the hideous possibilities inherent in some radical attempts to manipulate nature — instinctively give rise, and partly because of the use of the very word "engineering", with its implicit suggestion of man engineering man, and the fear it arouses of some of the fantasies of science fiction being enacted in real life. But we must consider the facts as dispassionately as we can; for the term "genetic engineering" is in fact used to cover two quite distinct categories of experimentation: first, the "selection and manipulation of sperm and ova to determine the genetic make-up of future offspring" — that is, many of the possibilities opened up by artificial insemination; and, secondly, "the manipulation of body cells which could be a way of substituting good genes for deleterious ones in a diseased person".[5] The first of these categories is already a very real possibility; and that must, I think, constitute our starting point. But the second opens up, in principle, almost limitless horizons, only some of which involve artificial insemination. In this sense, however, genetic engineering is still in the stage of basic research and future dreams, and there seems to be little possibility of any very spectacular developments in the foreseeable future.[6]

Artificial insemination of animals is, of course, already widespread, and today represents one of the major methods of economical and selective breeding. It has a long history, and there are reports of Arab tribes using it to inseminate horses as early as the fourteenth century.[7] Among human beings, on the other hand, artificial insemination was originally used to enable a childless couple to have a child which was unlikely to be conceived in the natural course of married life. Such artificial insemination by a husband (A.I.H.) was, it seems, first performed in London by a doctor named John Hunter about 1785[8] — although the whole subject of the artificial insemination of human beings does not appear to have attracted public attention at all widely until towards the end of the Second World War.[9] It can be used between husband and wife in cases where the husband, while not sterile, is less fertile than the average, or where the wife has some defect which militates against the husband's sperm penetrating far enough into her cervix. In most cases this simply involves the collection of a sufficient quantity of the husband's semen (which could well be obtained in the ordinary course of marital intercourse) and its injection into the wife's womb. Only in those cases in which the

[5] Cf. *Our Future Inheritance: Choice or Chance?* (A Study by a British Association Working Party), by Alan Jones and Walter F. Bodmer (OUP, London, 1974), p. 8.

[6] Ibid.

[7] Op. cit., p. 12.

[8] Cf. Glanville Williams, *The Sanctity of Life and the Criminal Law* (Faber and Faber, London, 1958), p. 110.

[9] Cf. *Our Future Inheritance*, p. 12.

husband, although fertile, is totally unable to have normal sexual intercourse with his wife, would it be necessary to resort to masturbation. But in some cases a couple's infertility may be caused by some more basic defect in the wife, and in such circumstances a much more complex operation may soon become possible; but at this point I had better quote my authority *verbatim.*

> Work is now going on in laboratories in various parts of the world to cure one particular form of the wife's infertility: that in which she has at least one healthy ovary and a healthy uterus, but the Fallopian tubes are either diseased or incapable of channelling the ova, or eggs, between the organs. This work has caused many a furore. In essence, the technique used, *in vitro* fertilisation, consists of removing the ova from the ovaries by a minor operation, using a hypodermic needle[10] inserted into the abdomen, and then fertilising these in a dish with the husband's semen. This procedure has been accomplished successfully and seven-day-old fertilised eggs have been grown outside the body. Up to the end of 1975, however, there had been no reported success in transferring the young embryo back to the womb for it to develop and grow naturally, a procedure which has been made to work in some animals. The chief problem is that the hormone balance of a woman who is a few days into pregnancy has in some way to be faithfully reproduced before the embryo can attach itself to the wall of the uterus. As yet this hormone balance is not sufficiently well understood.[11]

Self-evidently, however, precisely the same techniques can be used in a far more radical way. Where a couple is childless because the husband is totally infertile, and they earnestly desire children, it has already become fairly common — particularly in the U.S.A. — for the wife to receive artificial insemination by a donor (A.I.D.), that is, for sperm provided by someone other than her husband to be injected into her womb. This procedure is also, of course, sometimes adopted

[10] I am told that this is in fact done by means of a laparascope, which is a large hollow rod with a telescope inside it, and scarcely a needle!

[11] Cf. *Our Future Inheritance*, pp. 4 f.

where the problem is not the husband's infertility but a danger that he may pass on to his children some grave genetic defect with which he is himself afflicted, or which he is liable to transmit.

But at this point a wide variety of quite different possibilities opens up. To begin with, this method might well appeal to a woman who has an almost pathological longing for a child but an inhibition against marriage, or who has not met a man with whom a happy marriage seems possible. Yet again, it might open the door to the selective breeding of human beings in the manner of race horses, prize cattle or pedigree dogs; for it would be possible for a woman to be impregnated with the semen of a man of outstanding intellect, musical talent or athletic prowess. Naturally enough, this is no new idea, as is illustrated by the story of Bernard Shaw being approached by the dancer Isadora Duncan with the suggestion that they should have a child, since her body and his brain would make such a marvellous combination — to which Shaw promptly replied "Yes, madam; the snag would be if it had your brain and my body!" But the startling fact is that artificial insemination, combined with the resources of modern technology, would now make it possible for a contemporary Shakespeare, Einstein or Muhammed Ali to father children generations after his death.

Experiments in regard to the fertilisation of ova *in vitro* give rise to a number of yet further possibilities to which passing reference must also be made. As soon as the problem has been solved of how successfully to transfer to a human womb a young embryo which has been fertilised *in vitro*, for example, it would be possible for a woman who was liable to abort, had had a hysterectomy, feared the pain of childbirth or was unwilling to face months of pregnancy, to procure another woman to bear her child for her.[12] Fertilisation *in vitro*, moreover, would provide an opportunity, at some future date, for attempts to be made to treat some genetic defect more easily

[12] As I write, my attention has been called to an article in a newspaper stating that two surgeons in New York will, "within weeks", transplant a living embryo from the womb of one woman to that of a sterilised sister, who hoped to give birth to it in a perfectly normal way. This might, I suppose, be regarded as closely analagous to an organ transplantation, on the one hand, and to the adoption of a sister's baby, on the other.

than with an embryo *in utero*. It would also, presumably, be a prerequisite of human "cloning", if this is ever seriously contemplated. But this is a very technical subject in which, again, I feel I must quote my authority verbatim:

> To clone means to produce populations of cells or organisms from a single cell or common ancestor by the normal process of cell division. This bypasses the sexual processes which lead to a scrambling of the genetic material of both parents, and so the products of a clone (except for rare mutations) are genetically identical cells or individuals. Every time a plant is grown from a cutting the plant is, in effect, being cloned. But the application of the idea of cloning to animals is much more difficult. The possibility of doing so follows the much-heralded success of zoologists in producing identical clones of frogs . . . The procedure for cloning a frog is essentially simple. A cell is taken from a frog embryo in an early stage of development and its nucleus injected into an unfertilised egg which has had its cell nucleus removed or killed. In such circumstances, with selected species of amphibia, a significant proportion of the transplants develop into adults and the frogs which emerge are similar to the ones that grow from normally conceived embryos but their characteristics correspond to those of the animals from which the nucleus was removed. In 1971, zoologists at the University of Oxford transferred nuclei from cells grown in the laboratory (rather than from embryos) and succeeded in obtaining a few adult animals from such cell nuclei.[13]

But it must be noted that these cells were not synthetically produced, but taken from tadpoles — and that no success has yet been reported in the cloning of frogs with cells taken from adult frogs.

The question may well be asked why it has proved possible to clone amphibia but not, so far, mammals? The answer given is that the zoologists working on cloning think that there is

[13] *Our Future Inheritance*, pp. 112 f. But for the whole subject of embryo transfer, cf. a Ciba Foundation Symposium, *Law and Ethics of A.I.D. and Embryo Transfer* (Elsevier, Excerpta Medica, North Holland, 1973), pp. 11–17, 21, 37, 61, 73, 78, 85 and 93.

nothing inherently impossible in mammals being cloned, but the techniques necessary to achieve success have not yet been developed. "In essence the problem is that mammalian eggs are much smaller and more difficult to manipulate and the conditions of early development in mammals are much more difficult to control." But, however this may be, it may be stated that the cloning of humans — "with the rather frightening possibility . . . of producing armies of identical individuals with carefully planned genetic make-up" — is "still almost as remote a possibility as it has ever been".[14] It may, however, be observed in passing that this might possibly one day provide a method by which a scientist could contrive a virgin birth.

But it is time, I think, to turn from a mere specification of some of the possibilities already open (or which appear to be on the verge of opening) to medical science, and some of the ways in which the relevant techniques might be used, to a consideration of some of the objections (whether valid or invalid) which might be raised not only by a geneticist or a lawyer, for example, but also on moral, aesthetic or specifically Christian grounds. So let us deal first with the simplest case — artificial insemination by a husband.

To this there can, I think, be no objection whatever on either legal or genetic grounds — except, of course, for any genetic objections to the couple concerned having children by purely natural means. Nor would there seem to be any real moral objection, although many would feel considerable reluctance on aesthetic grounds to such an obviously "clinical" procedure, or might even regard it as repugnant to their innate sense of the dignity of man.[15] The Christian, moreover, will presumably remind himself that the procreation of children is not the primary purpose of either life or marriage. On the contrary, his basic ambition, whether single or married, should be to do the will of God and to grow in maturity and holiness; and he should not marry at all unless he believes that marriage will further this purpose. Again, a marriage may be all that God intends it to be without the advent of children — and the fact that they do not come may well be accepted, in some cases,

[14] Ibid.

[15] Cf. Note by Dean W. R. Matthews in *Artificial Human Insemination* (SPCK, London, 1948), p. 60.

as an indication of the divine will. If, however, one or both of the parties long for children, and if a medical examination indicates that a surgical operation would make that possible, there can be nothing intrinsically wrong in either of them submitting to such an operation. And much the same argument would, it seems to me, apply to A.I.H.

It was the firm conclusion of a commission of theologians, doctors, lawyers and others, appointed by the then Archbishop of Canterbury in 1945 to study the whole subject of artificial human insemination, that:

> 1. When the procedure can properly be described as "assisted insemination" (i.e. as a sequel to normal intercourse, or the attempt at it, between husband and wife), it may be justified.
>
> 2. When "assisted insemination" is inapplicable, or ineffective, other methods of artificial insemination with the husband's semen may be employed. Even if, for the insemination of a wife with her husband's semen, there is no practical alternative to masturbation by the husband, his act, being directed to the procreative end of the marriage, may be justifiable.[16]

These two findings (the first of which was unanimous, while the solitary dissentient from the second later withdrew his dissent) were subsequently re-examined and then presented on behalf of the Church to a departmental committee set up by the Government in 1958 to enquire into the same subject. And the same *moral* considerations would seem to me basically applicable, should medical research make this possible, to the fertilisation of the wife's ovum by the husband's sperm *in vitro*, and its subsequent implantation in the wife's womb — with the solitary *caveat* that no unacceptable risk of injury to the embryo is involved.

But when we turn to the question of artificial insemination by someone other than the husband (A.I.D.), however this may be effected, quite different considerations would seem to apply: genetic, legal, moral and specifically Christian. From the medical point of view, it is clearly important that all donors

[16] Ibid., p. 58.

should be suitably chosen, while the institution of a "sperm bank", for example, would open up the possibility of hundreds of babies being sired by the same man, with the subsequent risk that they might intermarry. But there are also exceedingly complex genetic problems — to say nothing of moral objections — in regard to any application to human beings of the sort of techniques of selective breeding commonly practised in regard to animals. To begin with, even if we knew what sort of persons we really wanted to breed, our knowledge of precisely how different genes interact on each other is still, I understand, fragmentary in the extreme. Human beings, as Dunstan insists, "cannot be bred for single characteristics, as calves can be either for beef or milk, or battery hens simply for eggs";[17] for the complex interactions of body, mind, disposition, moral fibre and social adaptability in human beings put them on an entirely different footing from animals in this respect. It seems to me distinctly doubtful, indeed, whether the human race would really benefit from an attempt to breed a disproportionate number of persons of outstanding ability, for society often finds it difficult enough to accommodate those eccentric men of genius it normally produces! The danger that someone might one day be tempted to go to the other extreme, moreover, and attempt to breed a strain of physically powerful morons for menial tasks, is too dreadful to contemplate. But to turn to a different, and much more imminent, possibility: if it should prove feasible before long to predetermine the sex of children before they are conceived (that is, by isolating the X and the Y bearing gametes in the male sperm before they have the opportunity to fertilise the female ovum), the imbalance of society which might result could have the most serious social, and even moral, consequences. In all these cases the differences between men and animals is obvious enough. Any experiments in the animal kingdom which led to unfortunate results could, for example, be remedied by the simple expedient of putting the animals concerned painlessly to death, while any imbalance between males and females would not depend on the predelictions of parents but on the considered demands of agricultural economy.

Unlike A.I.H., A.I.D. also gives rise to a considerable

[17] *The Artifice of Ethics* (SCM Press, London, 1974), p. 65.

number of legal problems. First, there is the question whether it does, or does not, constitute adultery according to the legal definition of that term. All the traditional definitions of adultery presuppose some act of sexual intercourse or "criminal conversation", as Glanville Williams observes;[18] and the lay mind would almost certainly concur. But this is not altogether conclusive in English law, since in a case in which two persons were in close bodily contact which did not actually involve intercourse, but did result in the woman's fecundation, this was held to constitute adultery. In Scotland, on the other hand, it was specifically decided in 1958 that A.I.D., even without the husband's consent, does not amount to adultery by the wife; and this decision would probably be accepted today as good law in England also.[19] Most people would probably agree with the view that a clear distinction between adultery and A.I.D. can be found in the fact that one is "done clandestinely to enjoy carnal pleasure", while the other is done "frankly to beget offspring without the emotional and physical enjoyment of coitus"; but others would, no doubt, argue that the legal essence of adultery "consists, not in the moral turpitude of the act of sexual intercourse, but in the voluntary surrender to another person of the reproductive powers or faculties".[20]

This question is still, I think, relevant in regard to divorce; for although the sole ground for divorce in this country is now the apparently irrevocable breakdown of marriage, a number of circumstances have been specified in the Divorce Reform Act, 1969, which constitute *prima facie* evidence of such breakdown — including adultery by one party which the other party regards as making the continuation of married life intolerable. So a strong argument could be made out for a wife's conception of a child by A.I.D. without her husband's consent being explicitly included in the relevant legislation — on a par with adultery — where the husband found this to be

[18] *The Sanctity of Life and the Criminal Law* (Faber and Faber, London, 1958), p. 119.

[19] Cf., respectively, *Russell* v. *Russell* [1924] A.C., at 721 (House of Lords), and *Maclennan* v. *Maclennan* [1958] Scots Law Times 12.

[20] Cf. *Artificial Human Insemination*, pp. 39 and 37. Cf. also Olive M. Stone, in the Ciba Foundation symposium mentioned above, p. 69, and Lord Kilbrandon's conclusion that *Russell* v. *Russell* was no longer of any importance in this context. (Ibid., p. 91).

intolerable. But it would, in my view, be quite unreasonable to regard the act of the donor as in itself providing *prima facie* evidence of the breakdown of *his* marriage. Again, problems regarding the child's maintenance and rights of inheritance would also arise. The question of maintenance would, presumably, depend in part on whether the husband had given his consent, in part on whether the child is in fact regarded as illegitimate, and in part on the husband's liability (or otherwise) to maintain any other illegitimate child born to his wife. It is distinctly possible, too, that the donor might be regarded as responsible for the maintenance of the child under the laws relevant to bastardy; in which case, seeing that a single donor might, as we have seen, have sired a hundred or more children, his theoretical liability would be somewhat daunting![21]

But is the child in fact to be regarded as illegitimate? There seems to be no doubt whatever, as the law now stands, that it is; and the only thing to protect him from this consequence, as Glanville Williams makes clear, is the presumption of legitimacy accorded to a child born to a married woman, which can be rebutted only by clear evidence that the husband *could not* have been the father.[22] But recent research has made it possible, I understand, for paternity to be absolutely excluded in such cases, should the suitable tests be authorised, on the grounds of the genetic differences between the child and its mother's husband. And there is a further legal point which carries obvious moral implications: namely, that the registration of a child begotten by A.I.D. as the child of the husband would be a criminal offence under section 4 of the Perjury Act, 1911.

It may be argued, however, that the legal problems are by no means insoluble. There are, of course, those who believe that

[21] This was recommended by the *Royal Commission on Human Artificial Insemination* (Cmd. 9678, para. 90), and the Departmental Committee on the same subject (Cmd. 1105, paras. 114–17). Lord Kilbrandon has asserted that the courts would give a divorce on these grounds even as the law now stands (cf. Ciba Symposium, p. 91).

[22] Cf. *Russell* v. *Russell*, at pp. 705–6. Also, for the possible responsibility of the donor for the maintenance of children begotten by A.I.D., cf. Glanville Williams, op. cit., pp. 116 f. But there are mutually conflicting decisions on the question of legitimacy in the U.S.A. (Cf. Ciba Symposium, p. 49).

A.I.D. should be prohibited, since "it is the function of law to prevent, and not to facilitate, fraud and deception";[23] while others believe that this would be impracticable and even undesirable, but would support legislation designed to control its use, to prescribe suitable conditions, to specify how a child so born should be registered, and to clarify the legal status he is to enjoy. A panel appointed by the British Medical Association in 1971, under the chairmanship of Sir John Peel, went so far as to recommend that A.I.D. should be made available, on a limited basis, within the National Health Service; that the definition of legitimacy should be expanded to include a child born of A.I.D. to which the husband of the mother had consented; and that for the purposes of registration of the birth of such a child the husband should be considered to be the father. But no action has in fact been taken; and it has been justly remarked that these recommendations "raise the question of how far legal fiction ought to go. Words become steadily more useless if they are stretched to mean what on their face they clearly do not mean."[24] But, whatever the law may decree, the moral objections to A.I.D. seem to me to be formidable.

It may, I think, be accepted as axiomatic that a wife who gives birth to a child by A.I.D. without her husband's consent has violated the essence of their marriage. I should not, myself, use the word "adultery" to describe what she has done, but it certainly constitutes an offence against the exclusive nature of her marriage vows. The moral problem is much more difficult, however, where the husband has given full and free consent — knowing, for example, that he could not, or should not, himself father a child. In such cases we must certainly try to understand, and sympathise with, the positive craving that some women — and a few men — have for children; with the fact that the wife may not merely long for a child in a way that, she feels, could be satisfied by adoption, but may have a deep desire actually to carry and give birth to her own baby; and

[23] Cf. *Artificial Human Insemination*, p. 42.

[24] *Our Future Inheritance*, p. 13. But this problem could be solved by Lord Kilbrandon's suggestion that the registration of births should provide for the registration of the name of the "father or accepting husband". (Cf. Ciba Symposium, p. 93).

that a husband who truly loves his wife may feel that he would prefer a child which is half hers, and may well resemble her, to a child which does not partake of the blood or genes of either of them. In addition, a husband who cannot himself father a child may suffer from a sense of personal inadequacy, failure or even guilt; and the marriage itself may be in jeopardy.

All the same, it seems to me that serious moral problems remain. First, I find it difficult to believe that a maternal instinct *cannot* be satisfied by the loving care and nurture of an adopted child, or that the months of pregnancy and hours of labour are quite so vital as this argument assumes — although this may be true of some women. Even so, no serious moralist can base a system of ethics on what a man or woman may desire, however passionately (sex outside marriage, for example, or homosexual sex); and it would seem to me that where a Christian couple cannot themselves beget a child, but long for a family, the obvious solution is to lavish their affection on some child who would otherwise be deprived of home, love and security. Admittedly, it is becoming increasingly difficult to find a suitable child for adoption; but, if this is not possible, I should have thought that Christians, at least, might accept the situation as coming from the hand of God — and it is also relevant to remember that the world is already suffering from over-population rather than under-population.

To reinforce this point I should like to quote in full a paragraph from the "Theological Statement" in the Report of the Archbishop's Commission:[25]

> The strongest plea put forward in defence of A.I.D. is the plight of those married women who long for children but whose husbands are sterile. Repeatedly it is said — by such women and on their behalf — that what they need and desire above all else it to have "a child of their own". "Nothing else in the world will satisfy them." We need not affirm once more the profound compassion which this frustration must evoke; but our concern at this point is to answer the question whether such a desire — one which impels a wife to become the mother of a child who is not her husband's, but

[25] *Artificial Human Insemination*, p. 50.

> some other and unknown father's — is in strict truth *inordinate*: one which exceeds the proper bounds of desire. There are some whose sexual impulses might be similarly described; but that is no defence in a court of law, nor, we suppose, in the eyes of the public at large. We are all, without distinction, required to restrain our desires, however imperious. On what rational ground is it urged that while *sexual* desires ought not to be indulged at will, *parental* desires may be? And are the results of indulgence in the latter likely to be quite different, in their total effect on the personality, from those which are known to follow in the former? If we persuade ourselves that because we want a thing so much it must be right for us to have it, do we not thereby reject in principle, though perhaps unwittingly, the very idea of limitation, acceptance, of a given natural order and social frame — in a word, of the creatureliness of man?[26]

There are also a plethora of other possible problems. To begin with, it seems distinctly possible that a husband's attitude to a child begotten in this way, even with his consent, might change with the years. If, moreoever, the child developed marked defects of body, mind or character — or, indeed, elements of genius — this feeling of estrangement might well be accentuated. In any quarrel or disagreement about him, again, the wife might at times be greatly tempted to claim that he was uniquely hers, and the husband to disclaim any responsibility — by contrast with an adopted child, in regard to whom husband and wife are on an equal footing.[27] It is, moreoever, usually considered wise to tell an adopted child at an early age, simply and naturally, the fact of his adoption; but this would be much more difficult in the case of a child born by A.I.D. — although it might become imperative, when he became adult and contemplated marriage, should he know that the man he believed to be his father suffered from some serious genetic defect. It is distinctly possible, indeed, that the secrecy

[26] There is also the exceedingly difficult problem of whether a very strong desire is *physiological* or *pathological*. Where it is pathological, the moral answer would, presumably, be to seek cure rather than fulfilment.

[27] But it is only fair to observe that I have discussed these points with the consultant who has more experience than anyone else in this country in such cases, and he assures me that these fears have little foundation.

which necessarily cloaks the whole procedure — together, perhaps, with the mother's growing appreciation of the equivocal status of her child and a lurking sense of moral guilt — might undo the good his birth was designed to effect. And the moral arguments against a single woman satisfying her maternal instincts by this means, while somewhat different, are at least equally strong — to say nothing of the psychological and other considerations which normally militate against a single woman adopting a child. Nor am I at all happy about what is required of the donor.[28] As for the basic morality of one woman bearing another woman's child for her — by the implantation of a fertilised egg in her womb — this would, I think, be dependent in part on the motives of both the biological mother and the foster mother; for there would be a manifest difference between the case where a foster mother hired her body for the convenience of someone who merely wished to avoid the cost of pregnancy and labour, and one who acted self-sacrificially for a friend who would have loved to bear the child herself. Even so, I would myself regard the procedure as distinctly questionable. And the very idea of the selective breeding of human beings in the manner of race horses or prize cattle seems to me to constitute a fundamental degradation of man made in the image of God.

It is true that Julian Huxley, for example, hailed with joy the fact that "the perfection of birth-control", and "the still more recent technique of artificial insemination", had "opened up new horizons" for men and women "to consummate the sexual function with those they love, but to fulfil the reproductive function with those whom on perhaps quite other grounds they admire". But the Christian must, I think, regard this as an obvious instance of man putting asunder what God has joined together. It is not for nothing that Huxley continues: "But the opportunity cannot yet be grasped. It is first necessary to overcome the bitter opposition to it on dogmatic theological and moral grounds, and the widespread popular shrinking from it, based on vague but powerful feelings, on the ground that it is unnatural."[29] But this brings us back to the principle of what we termed the "creation ordinance" of the basic institution of

[28] Cf. Dunstan, in the Ciba Symposium, pp. 52 f.
[29] *The Uniqueness of Man*, pp. 78 f.

marriage and family relationships; and just as sex, for the Christian, must (for this reason among others) be confined to marriage, so too, it seems to me, should the conception of a child. And where this is impossible, then the obvious alternative would appear to be adoption; for A.I.D., it has been justly remarked, represents "the acme of the depersonalisation of sex... It extracts procreation entirely from the nexus of human relationships." Incidentally, to argue — as I do in these lectures — that we should not follow those who seem to suggest that what we call "Nature" represents all that God intended it to be, and that we therefore touch it at our peril, is not to deny that we can still say that some things (such as homosexual relations, for example) are "contrary to nature", or to those creation ordinances, or "Maker's Directions", which are revealed in the Bible and written, in some measure, on all men's hearts.

Early in this lecture I said that the term genetic engineering is used in regard to at least two distinct categories of experimentation: first, "the selection and manipulation of sperm and ova to determine the genetic make-up of future offspring"; and, secondly, "the manipulation of body cells which could be a way of substituting good genes for deleterious ones in a diseased person". So we must now turn to the second part of our subject, and there can be no question what an urgent problem this represents. "Perinatal mortality in Britain has been brought down from more than 200 per 1,000 births at the turn of the century to about 20 per 1,000 now, but the number of deaths from congenital malformation has remained constant over the period at 5 per 1,000." In other words — as Professor R. J. Berry insists — "the mortality due to a group of diseases with a high genetical component 'has increased from a negligible two per cent to a substantial 25 per cent... Six per cent of the population have a readily recognisable genetical defect at birth; and twenty-five per cent of children admitted to hospital are there because of some inherited anomaly.' "[30] So, as deaths from environmental causes (pneumonia, tuberculosis, etc.) decrease, the proportion of deaths from genetical disease becomes increasingly significant. It is partly for this reason that

[30] Cf. article on "Genetical Engineering", in *The Christian Graduate* (March, 1973).

G. Rattray Taylor has expressed the view that "the power to interfere in the processes of heredity is the most serious of all the human problems raised by biological research".[31]

Reference has already been made to one way in which an attempt might be made to tackle this problem: namely, selective breeding. But this would not really help, even if we knew what we wanted to breed for, partly because so little is known about the principles which govern congenital malformation, and partly because of the extreme difficulty of restricting the liberty of individuals to reproduce as and when they wish[32] — for in selective breeding the negative side is quite as important as the positive. In our present state of knowledge it is, indeed, much easier to think in terms of contraception, the sterilisation of the genetically defective, and the abortion of embryos which can be shown to be diseased, than in terms of cure: but any discussion of this negative aspect of the problem must be reserved for my next lecture. In this we are concerned exclusively with the positive aspect of attempting to cure genetic diseases by modifying the gene itself. In any such attempt, however, the approach to the problem depends on whether the object in view is to cure the disease in some individual or in his progeny, for "in the former case it is the cells of the diseased tissues or organ that must be changed, while in the latter it is the sperm or the ovum". And though the techniques of biology may be similar in these two situations, their goals are quite different, for "curing the cells which cause the disease — gene therapy — has no further genetic implications".[33] If, for example, a patient suffering from a disease which takes the form of a malfunction of the cells of the liver could be treated by the insertion into his liver of normal genes, the disease might be cured; but the genes the patient would pass on to his offspring would still be abnormal.[34] To cure sperm or ovum cells would have far wider implications, for in this way an hereditary cure would be effected and the person concerned would pass on normal genes to his or her offspring. I understand, however, that "while the prospect of manipulating cells other than the sperm or ova are, at least in a number of cases,

[31] *The Biological Time Bomb* (Panther, London, 1969).
[32] *Our Future Inheritance*, p. 107.
[33] Op. cit., p. 109. [34] Cf. op cit., p. 106.

not too remote, any prospect of curing the sperm or ova, especially on a large scale, is still a long way off".[35]

But what, it may be asked, are the possibilities for the future? Human beings are estimated to have about 50,000 different genes, which may be defined as an inherited factor carried on the chromosomes in the nucleus of each cell; and one of the most important recent advances in biology has been the discovery of the precise chemical nature of the gene. This is "now understood to be a molecule of the substance known as deoxyribonucleic acid, or DNA for short".[36] Molecules of DNA have, moreover, been synthesised in a test-tube, and the goal of the modern genetic engineer, we are told, "might be described as the complete synthesis of a set of genes", which "would be followed by the incorporation of these genes into a cell in such a way that it would divide and express precisely the properties of the synthesised set of genes. Given that the 'language' of the gene is understood, organisms of desired types could then by synthesised at will." This would certainly be a daunting prospect if applied to man; so it may be providential that the scale of the problem is positively monumental, since the "DNA sequences that have so far been synthesised are, at most, a few tens of units long, as compared with the total DNA sequence of man, which has some 3,000 million units in it". First, therefore, it would be necessary to synthesise one or only a few of the relevant genes, and then to put the gene in its appropriate place, so that it could replace some defective gene. But "apart from the extremely difficult task of synthesising even just one gene (on average likely to be several hundred to a thousand units long), the problem of getting this to its right place in the genetic material of a cell is formidable".[37]

There is also the question of which cells to use. "Many sorts of cells can now", it seems, "be grown in culture just as if they were microbes", and "one of the hopes is to manipulate these in culture and then put them back into the body having cured whatever defect they carried". Research along these lines is, in itself, "a natural consequence of the geneticist's desire to obtain a better understanding of basic cellular mechanisms"

[35] Op. cit., p. 109. [36] Op. cit., p. 107.
[37] Op. cit., pp. 107–9.

which underlie the causes of many diseases such as cancer. One comparatively recent achievement has been the discovery of the process known as "bacterial transformation". This, it seems, "is a form of 'genetic crossing' in which naked DNA, the chemical substance, is one of the parents and is mixed with whole bacteria as the other parent" — and in the late 1940s "genetic systems for making such crosses were discovered in both bacteria and in the viruses that attacked them". It is the ability to manipulate the genetic material of bacteria and their viruses in a controlled way that has "opened up many of the possibilities of molecular biology in the past 20 years".[38] But two major caveats remain. First, we are told that "while in principle it might be possible to develop similar systems for human cells, the technical difficulties seem to be almost overwhelming". Secondly there is the danger that experimental work in genetics might become a public health hazard through the creation of new strains of virulent viruses and bacteria against which there may be no antidote.[39] So the Government is appointing a working party to draw up a code of practice — based, in essence, on a report by Lord Ashby on the potential benefits and potential hazards of genetic manipulation, or the transfer of genetic material between micro-organisms to create new strains.

Other possibilities also exist. One of these is based on the fact that cells of quite different types can "be mated together just like bacteria". In this way it is even possible "to cross mouse cells with human cells and make a hybrid cell containing both human and mouse genetic material" — but only, I understand, for purely experimental purposes! The ultimate objective is that the secret may one day be discovered of how one or more genes could be transferred from a normal cell to an abnormal cell; but there would be a danger that unwanted genetic material might be wrongly transferred, that the genetic balance of the receiving cell might be upset, or that a transferred cell might prove to be malignant. Another possible approach would be to use "a virus attached to a necessary gene . . . as a vehicle for delivering a gene to the cells of a patient suffering from a hereditary defect" and so effect a cure.

[38] Op. cit., p. 110.
[39] Cf. *The Times*, August 7, 1975.

The claim has recently been made that "results suggest that it might be possible to introduce a selected bacterial gene into human cells *in vivo*, using phage (bacterial virus) as a vehicle, and significantly to alter a specific metabolic pathway". And yet another method would be by "direct gene surgery by the manipulation of chromosomes by laser beams, etc." But this would involve the danger that other genes might be damaged in the process.[40]

Finally, in so far as this lecture is concerned, what should be our attitude to all this? In a fascinating passage Dunstan summarises two different approaches: that of the Protestant Professor Paul Ramsey of Princeton, and the Roman Catholic Father Bernard Häring of Rome and Bologna. Contrary to what one might expect, Ramsey's attitude is much the more negative, for he is distinctly uneasy not only about the work which geneticists and embryologists are now doing, but — still more — the procedures which might become possible at some future date. "He defends man as an end in himself, whose interests may not be subordinated, without consent, to any interest outside himself, particularly to the abstract ends of 'the good of humanity', 'the future of the human race', or 'the progress of medical science'. He would allow no proxy consent, especially that of a parent for a child born or unborn, to a procedure not intended directly for the good of that particular patient but only or even primarily for 'research'. He would carry these defences back to the very beginning of life, so that he would reject any experiment in 'genetic engineering', for example the introducing of a virus to modify a defective gene (if it became possible) unless it were already *known* that there were no hazards at all."[41]

Haring, on the other hand, while he shares Ramsey's general principles, is much more liberal in his approach. He believes that "biological processes normally serve the good of the total human person": when they do, they are to be followed and further developed, but when they do not, it is a human duty to experiment and search out the means to improve them — for "responsibility for the world in which men are destined to live together is one of the most sacrosanct duties of man, infinitely

[40] Cf. Berry, loc. cit., p. 6.
[41] Quoted from *The Artifice of Ethics*, pp. 59 f.

more sacred than biological processes", and "each intervention or medical provision that helps or enhances the wholeness of the human person is right".[42] So "today's physician is less a 'servant of nature' than a creative manipulator within the forces of 'nature' ". Even man's true "nature", in its empirical reality, has yet to be realised, Häring believes (which seems to me, for reasons to which I have already referred distinctly questionable), and medical research and intervention are means, within their relevant sphere, towards bringing it into being. But this still leaves him "the task of deciding which particular interventions are ordered to this end, and what treatment is licit or illicit for any particular patient"[43] — and even opens up the terrifying possibility that decisions about the manipulation of genes, or future generations of men and women, might in practice be made by some scientific boffin or political schemer.

Now this, of course, brings us back full circle to the problem with which we began. "Nature", as such, is not inviolable, for nature itself has been affected by cosmic sin. Nor is there anything wrong with man acting as a "creative manipulator", provided that he always remembers that he is exercising only a delegated authority, that he is himself prone to both sin and error, and that he must do his utmost never to act contrary to the essence or implications of the revealed will of the Creator himself — judged, *inter alia*, in terms of the reverence due to man made in the image of God, of the family structure in which he has been placed, and of his duty of service to the human race as a whole.

[42] Op. cit., p. 61. Cf. Bernard Häring, *Medical Ethics* (St Paul Publications, 1972), pp. 53–63.
[43] *The Artifice of Ethics*, p. 61.

3 Birth Control, Sterilisation and Abortion

IN MY LAST lecture my subject was Artificial Insemination and Genetic Engineering. In other words we were largely concerned with what may be termed the positive aspect of eugenics — that is, the science of trying to improve the hereditary qualities of future generations of men and women. But an equally important side of eugenics is the negative approach: the attempt to prevent the birth of children who, for one reason or another, ought not to be born. And it is this approach — whether by means of contraception, sterilisation or abortion — which will, in part, now engage our attention.

Now it is immediately obvious, I think, that each of these methods of preventing the conception or birth of children can be used for a variety of different purposes and from a number of different motives — some of which we considered last week in the context of the positive, rather than the negative, approach. Much has been heard of late, for example, about one aspect of the broad concept of what may be termed "selective breeding": and I do not now refer to the attempt to breed super-athletes (as may, for all we know, be already on the way), nor the dream of a Hitler to breed a master race, but to the disproportion in the size of the families born to parents who are able to give their children a good start in life — whether judged by their intellectual and cultural background, their financial circumstances or the geographical area in which they live — compared with those born to parents who are ignorant, poor, and at times feckless. It was to this problem that Sir Keith Joseph referred in such unfortunate terms when he spoke of the size of the families of those in the "fourth and fifth socio-economic class". But this, naturally enough, aroused a positive storm of protest — and should, in any case, be countered by Professor J. B. S. Haldane's quip that "If the unscrupulous become rich and the poor have more children, at any rate we should expect some moral improvement!"

Yet, however we phrase it, the problem remains — particularly in the "Third World". When a middle-class Englishman speaks like this he may easily be written off as a racist, or one who wishes to maintain the dominance of the West; but it is the Governments of many of the countries concerned which are most acutely conscious of the dilemma. Where can they find the *food* — or, in some cases, even the space — for the children who are being born in such profusion? How can they provide them with education, health services and eventual employment? But what are the alternatives? Much, no doubt, can still be done to increase food production by scientific means; and a great deal more should certainly be done to promote a more equitable distribution of the food which is being produced, and to curb the greed of the affluent West in favour of those in other countries who are starving or undernourished. But there are limits to what the earth can produce and what man can be induced — or even forced — to do for others. Every advance in medicine, moreover, serves to add to the problem, for the lives of millions of people who would previously have been wiped out by smallpox, cholera or tuberculosis are now being saved. As a result not only individuals but Governments are looking with new eyes at the ethics and practicability of birth control, sterilisation and abortion — whether voluntary or even, perhaps, by compulsion.

Again, we meet precisely the same problem in the attempt to control genetic defects. In my last lecture I referred to the ever-rising proportion of diseases which can be regarded as stemming from hereditary rather than environmental factors. So should A and B marry and have children, or should C have children at all? But, again, what is the alternative? Can society realistically demand that in all such cases A and B should not marry, or that C should observe complete continence? Or are we, in some circumstances, to encourage voluntary sterilisation; and is compulsory sterilisation ever justified? Then supposing a baby is in fact conceived which, it is feared, may suffer from serious genetic defects, what is to be done about it? It is becoming increasingly possible to detect such defects, in some cases, at an early stage in pregnancy and for a doctor to recommend abortion. But is this right — and, if so, in what cases? How serious must the defect be? And who is to make the decision?

It is equally obvious, moreover, that the same means may be used in circumstances which are far less morally defensible. Just as some married couples long for a child which is denied them, others wish to avoid — sometimes for purely selfish reasons — the physical and financial cost of pregnancy, childbirth and the care and upbringing of children. But by what criteria should Christian couples decide how many children they should have, when they should have them, and what method they should adopt to avoid an unintended pregnancy? And supposing, none the less, conception does take place, in what circumstances — if any — is abortion legitimate for the Christian? Again, what attitude should the Christian take to those who will not confine their sexual activities to married life? Are contraceptives and contraceptive devices to be made readily available to all who desire this, regardless of their age? Many would maintain that such an attitude tends to reduce moral standards to nothing higher than "Thou shalt not beget an unwanted child", while others would reply that the alternative — the conception of "unwanted" children — is too high a price to pay for an attempt to impose an enforced morality. But is there a middle course; and is abortion to be regarded as the ultimate remedy — whether for unsuccessful genetic experimentation or failures in contraception?

This means that we must approach each part of our subject in this lecture from several points of view. First, what attitude should we take to contraception (or different methods of contraception) *per se*; and in what circumstances is it ethically right — or even perhaps a moral *obligation* — to resort to it? Secondly, is voluntary sterilisation ever right; should parents sometimes be permitted to consent to the sterilisation of their minor children; and may the State ever legitimately interpose by way of compulsory sterilisation? Thirdly, is abortion ever permissible, and in what circumstances? Does it make any *ethical* difference how early or late in pregnancy it takes place? Again, who is to make the decision: the mother alone, or the mother and the doctor, or some committee? And supposing they disagree? Can a doctor be held liable for refusing to perform an abortion which ought to have been permitted; is a doctor who is totally inflexible on this subject to be excluded from certain appointments; and can pressure be brought to bear on a mother who wishes to insist on bringing into the world a seriously defective child?

First, then, the question of birth control, in regard to which Christian opinion has, in general, changed so dramatically in the last few years. For centuries the Christian conscience in this matter was largely formulated by the views of St Augustine and St Thomas Aquinas. In his *Marriage and Concupiscence* St Augustine laid down the principle that "It is one thing not to lie except with the sole will of generating: this has no fault. It is another to seek the pleasure of the flesh in lying, although within the limits of marriage: this has venial fault. I am supposing that then, although you are not lying for the sake of procreating offspring, you are not for the sake of lust obstructing their procreation by an evil prayer or an evil deed. Those who do this, although they are called husband and wife, are not; nor do they retain any reality of marriage, but with a respectable name cover a shame."[1] And this same attitude was declared by St Thomas to be inherent in Natural Law, by means of which man, as a rational creature, can perceive that the purpose of sex is the continuance of the species, so it is wrong to use it for any other purpose. And he "re-inforced this general and abstract view of the purpose of sexuality", as Mr St John-Stevas tells us, "with a very particular but equally abstract notion about the nature of male semen". This, St Thomas insisted, "is necessary for generation, and exists for the purpose of continuing the race"; so it "has a special significance, and any misuse of it is gravely

[1] Cf. Norman St John-Stevas, *The Agonising Choice* (Eyre and Spottiswoode, London, 1971), p. 67.

sinful". Thus with remorseless logic — but, I may venture to remark, with an almost total lack of common sense or practical morality — he went "so far as to maintain that in this class of sin only murder is worse, because murder destroys actual human nature, whereas misuse of seed destroys human nature in potential", and "that masturbation is worse than fornication or adultery, since in the latter, whatever the social effects, generation is not *ipso facto* excluded". For him, then, the all-important point was the purpose of insemination — or, indeed, of intercourse: the procreation of children. So its intentional and objective exclusion was a mortal sin, and "to avoid venial sin in sexual union there must be at least some subjective intention of procreation".[2]

But this seems to me remote from the teaching of the New Testament. It is true that there was a strong tendency towards asceticism in the early Church; but apostolic teaching on the subject of marriage insists: "Let marriage be held in honour among all, and let the marriage bed be undefiled;[3] for God will judge the immoral and adulterous." Again: "The body is not meant for immorality, but for the Lord, and the Lord for the body . . . Do you not know that your bodies are members of Christ? . . . Shun immorality. Every other sin which a man commits is outside the body; but the immoral man sins against his own body. Do you not know that your body is a temple of the Holy Spirit within you, which you have from God? You are not your own; you were bought with a price. So glorify God in your body."[4] Then, in the next chapter, St Paul writes ("by way of concession, not of command") that "The husband should give to his wife her conjugal rights, and likewise the wife to her husband. For the wife does not rule over her own body, but the husband does; likewise the husband does not rule over his own body, but the wife does. Do not refuse one another except perhaps by agreement for a season that you may devote yourselves to prayer; but then come together again, lest Satan tempt you through lack of self-control."[5] Married

[2] Op. cit., pp. 69 f.
[3] Hebrews 13:4. Or these two sentences may be translated as statements of fact rather than exhortations.
[4] 1 Corinthians 6:13, 15 and 18–20.
[5] 1 Corinthians 7:2–6.

couples, moreover, are exhorted always to behave "as heirs together of the grace of life, that your prayers be not hindered".[6]

Now surely the essence of this teaching is that sex outside marriage is sinful *per se* and comes under the judgment of God, and sexual indulgence, even inside the marriage bond, is unworthy of those whose bodies are "members of Christ"; but that due attention to the bodily needs of a marriage partner is incumbent on a Christian husband or wife, and the fundamental purpose of marriage is for mutual help in life, holiness and service, rather than the procreation of children. This does not mean, of course, that a Christian couple should be guided purely by their own whims or convenience in regard to whether they should have a family, and how large that family should be; for in this, as in all else, they should seek to know and do the will of God. But it runs on a completely different plane from the dicta of St Augustine and St Thomas.

So the revolution in Christian thinking in the last fifty years seems to me to represent, from one point of view, a return to New Testament teaching, although it has certainly been influenced by secular attitudes and by the problem of overpopulation. But that there *has* been such a revolution can be clearly seen from the pronouncements of a series of Lambeth Conferences. In 1920, for example, the Conference "while declining to lay down rules which will meet the needs of every abnormal case", uttered "an emphatic warning against the use of unnatural means for the avoidance of conception". "In opposition to the teaching which, under the name of science and religion, encourages married people in the deliberate cultivation of sexual union as an end in itself," the bishops insisted, "we steadfastly uphold what must always be regarded as the governing considerations of Christian marriage. One is the primary purpose for which marriage exists, namely the continuance of the race through the gift and heritage of children; the other is the paramount importance in married life of deliberate and thoughtful self-control."[7] The Conference of 1930, moreover, again stated that the primary purpose of marriage was

[6] 1 Peter 4:7. [7] Resolution No. 68.

the procreation of children, but went on to declare (by a majority vote) that:

> Where there is a clearly felt moral obligation to limit or avoid parenthood, the method must be decided on Christian principles. The primary and obvious method is complete abstinence from intercourse (as far as may be necessary) in a life of discipline and self-control lived in the power of the Holy Spirit. Nevertheless, in those cases where there is such a clearly-felt morally sound reason for avoiding complete abstinence, the Conference agrees that other methods may be used, provided this is done in the light of the same Christian principles. The Conference records its strong condemnation of the use of any methods of conception-control from motives of selfishness, luxury, or mere convenience.[8]

But the Conference of 1958 went a good deal further, and gave unanimous approval to a Resolution stating that:

> The Conference believes that the responsibility for deciding upon the number and frequency of children has been laid by God upon the consciences of parents everywhere: that this planning, in such ways as are mutually acceptable to husband and wife in Christian conscience, is a right and important factor in Christian family life and should be the result of positive choice before God. Such responsible parenthood, built on obedience to all the duties of marriage, requires a wise stewardship of the resources and abilities of the family as well as a thoughtful consideration of the varying population needs and problems of society and the claims of future generations.[9]

And a study group appointed by the World Council of Churches which met in 1959 at Mansfield College, Oxford, reached similar conclusions in their Reports on "Responsible Parenthood and the Population Problem", with the only dissentient voice coming from the Orthodox Church.[10] Even in

[8] Resolutions No. 13, on the purpose of marriage, and No. 15, on contraception.

[9] Resolution No. 115.

[10] Cf. St John-Stevas, op. cit., p. 75.

the Roman Catholic Church, moreover, the ramparts of rigorism are today being defended with ever-increasing difficulty.

But it would be wrong — as we have noted in passing — to regard this revolutionary change of attitude as a mere capitulation to secular views, undeniable though it is that Christian thought is by no means immune from secular influence. It can be attributed, in the main, to a compound of three different factors: the obvious danger of over-population in a world where numbers are today doubled in little more than thirty years — a factor which no responsible moralist can ignore; the advance in knowledge which makes it possible — and therefore, as many would argue, mandatory — to control a situation in which "human numbers are pressing against human values";[11] and a new approach to the purpose, and so the theology, of marriage.

The problem remains, however, whether all methods of birth control are equally permissible. This is a complex subject on which I am certainly not qualified to go into details; but I must, I suppose, make a few remarks of a very general nature. It will have been noted, for example, that the Lambeth Conference of 1920 specifically referred to "unnatural means for the avoidance of conception"; and this phrase was probably intended not only to exclude what the Conference of 1930 termed the "primary and obvious method" of abstinence from intercourse, but also the restriction of intercourse to the so-called "safe period", during which conception is unlikely. It is difficult, however, to make out a convincing case for this as the only permissible method of birth control. And it should, I think, be mentioned in passing that the near obsession with the story of Onan in Genesis 38:8–10 which at one time underlay much Christian thought about contraception was almost certainly misplaced; for however undesirable *coitus interruptus* may be on psychological grounds, the judgment on Onan does not seem to have had anything whatever to do with contraception, but was presumably a punishment for his fundamental refusal — and cynical misuse — of the levirate law, in an attempt to secure the privileges of the first-born for his own children.

When, moreover, we turn to what may be called "artificial"

[11] Cf. R. J. Berry, in "Genetical Engineering".

methods of birth control, any ethical distinction there may be seems to lie not so much in the method *per se* as in the way in which it operates: that is, whether it prevents the fertilisation of the ovum altogether, whether it prevents a fertilised ovum from becoming attached to the wall of the womb, or whether it destroys the embryo even after what is termed nidation. The classical method, which has today won such wide-spread approval, is of course the first: some method of preventing the meeting of sperm and ovum, which can be done by a variety of mechanical devices or chemical prescriptions. And while there may be a number of considerations (physical, psychological or otherwise) which incline some moralists to prefer one method to another,[12] the fundamental fact is that in all cases it can be stated that fertilisation has been prevented, and thus the inception of life has been frustrated rather than destroyed.

But there are other methods of birth-control in regard to which it is, it seems, exceedingly difficult to say whether they prevent fertilisation altogether or prevent an ovum which has, in fact, been fertilised from becoming attached to the wall of the womb. This is true of certain hormones and is particularly applicable to the copper-impregnated coil so widely used in parts of the Third World. In some cases, it seems, this inhibits implantation, but in others may well have a "potent inhibitory effect on sperm mobility".[13] So in the latter case many Christians would accept its use without any qualms of conscience, while in the former they would feel that a totally different ethical principle is involved. But is this really a tenable view?

Mr R. F. R. Gardner, in his Rendle Short lecture for 1973, argues that it certainly involves a number of difficulties. To begin with, there is the basic uncertainty — to which reference has just been made — as to how it does in fact work; to which the answer seems to be that this differs from case to case. But the basic difficulty is a severely practical one: that in many

[12] It always seems to me, for example, that a pill which prevents ovulation, and thus completely upsets the whole rhythm of a woman's biological processes, should be regarded as one of the most stringent of methods, however convenient it may be.

[13] Cf. "Moral Dilemmas in Contraceptive Developments", R. F. R. Gardner (published by the Christian Medical Fellowship, London, 1973), p. 7.

developing countries "the current sophisticated methods popular in the West are just not being used widely enough to stem the pandemic of births . . . So long as doctors and nurses must supervise the methods they are too expensive. Their continued use requires a larger degree of commitment and discipline than the majority of the population at risk are willing to give."[14] And the fact remains that, in Gardner's words, the Intra-Uterine Device is the standard method of birth control in those countries of the world where the need for it is most urgent.

> At least 12 million women are wearing them today. It is cheap, it can be easily inserted . . . recall visits are required infrequently and therefore do not require expensive resources . . . When one writes to missionary doctors actually working in these lands — as I have done — and asks whether in view of the possible implications they will continue to use the I.U.D. they reply that they *must*. They all say this, whether they work in India, Ethiopia, Kenya, Burundi, Rwanda, Tanzania or Botswana. Today, therefore, we have to face the fact that some of us as Christian doctors, in using I.U.D.s, are *already* inextricably involved in post-conceptual birth control.[15]

But a third, and still more drastic, method of birth control may soon be discovered: namely a "non-toxic and completely effective substance or method which, when self-administered on a single occasion, would ensure the non-pregnant state at completion of a monthly cycle"; and Gardner informs us that the United States Agency for International Development had, in 1973, already allocated more than $10 million for research into such a method. At the time of his lecture the prostaglandins appeared the most likely candidates for this role; and these, it seems, are already commercially available and are being used in obstetric units for the induction of labour. But there are other possibilities. "What we Christians have to decide," Mr Gardner insists, "is whether we are going to accept and use these agents, or whether this is a place to stand and fight. History is littered with instances where Christians

[14] Op. cit., p. 7. [15] Op. cit., pp. 9 and 10.

have fought needless battles over non-essentials. There have been others, however, where matters of grave importance have gone almost by default. If there is to be a battle on this matter, let it be on ground of our choosing."[16]

Now it is obvious that in any method of post-conceptual birth control, and still more post-nidatory birth control, the line between contraception and abortion becomes blurred. It will be convenient, therefore, to postpone any ethical discussion of these methods until we tackle the subject of abortion as a whole, and the exceedingly difficult question of when an embryo can be regarded as becoming in any real sense a human being or, in the language which is so often used, as possessed of personality, soul, spirit[17] or the capacity for human relationships. But before we turn to that most complex and fascinating subject we must deal briefly with the ethical problem of the provision of contraceptive prescriptions and advice for the unmarried, including the immature, and the whole question of sterilisation, whether voluntary or compulsory.

Contraceptive advice for those who are unmarried, and particularly for the young, presents the Christian with an acute problem. There can be no doubt whatever that the Bible repeatedly condemns fornication; that Christians are commanded to have nothing to do with it and not even to keep company with those who indulge in it; and that unrepentant fornicators are specifically mentioned as among those shut out of God's presence.[18] So should a Christian doctor in any sense "aid and abet such activity by prescribing in these circumstances contraceptive agents?" To many Christians the answer to this question will be an unequivocal "No"; and there can, I think, be only one ground on which they might come to the opposite view: namely, the doctrine of the lesser of two evils. If, for example, a Christian doctor knows that one of his patients is at risk, and if he has done all he legitimately can to dissuade her from her manner of life, which is the lesser evil: contraceptive advice, or the danger of an unwanted pregnancy and possible demands for abortion? This is an exceedingly uncongenial field for a Christian physician; but there may well

[16] Op. cit., p. 8. [17] Cf. pp. 18ff. above.

[18] Cf. Gardner, op. cit., p. 19 and Ephesians 5:3, 1 Corinthians 5:9, Revelation 22:15, etc.

be cases where he feels he cannot withhold advice, or must at least provide information as to where such advice can be obtained.

We now turn from contraception to sterilisation, which may be defined as depriving the person concerned of the capacity for reproduction. This means that what some forms of oral contraception do for a limited period sterilisation does permanently — except that the surgical operation by which sterilisation is effected is such that it can, in some cases, be reversed. It is important, however, not to confuse sterilisation with castration; for the latter is a much more radical operation which normally deprives a man of all sexual potency and has a marked effect on a number of secondary sexual characteristics, whereas sterilisation has no physical effect whatever except to prevent procreation. Castration has, indeed, sometimes been self-inflicted by those who gave a literal interpretation to our Lord's reference to those who "made themselves eunuchs for the kingdom of heaven's sake";[19] but this practice was declared by the church at an early date to be wrong, both as a matter of ethics and exegesis, and was also found to be spiritually ineffective in eliminating sexual fantasies. The operation was also, of course, performed on others to provide eunuchs to supervise or serve in Oriental *ḥarīms*,[20] in the belief that it deprived its victims of sexual potency as well as desire (although even this occasionally, it seems, proved to be illusory). But, however this may be, we are concerned only with sterilisation — which can be effected by a very minor operation in the case of a man and a rather more radical one in the case of a woman.

Like contraception, sterilisation is used today — although more rarely — for three purposes: for population control, for genetic reasons, and for convenience. It can also, of course, be effected for purely therapeutic reasons — that is, for the physical health of the patient concerned. As such it is usually, perhaps, a by-product of an operation performed for some other reason, as in the case of a hysterectomy made necessary by some defect of the womb; but in other cases the therapeutic

[19] Matthew 19:12.

[20] Or, indeed, at one time to provide male sopranos for the papal choir, but this was later forbidden by the Pope.

reasons and "convenience" shade into each other — e.g. where there is some therapeutic reason why the woman should not have another pregnancy. Unlike contraception, however, sterilisation can be compulsory as well as voluntary.

It is primarily in the context of population control, however, that sterilisation is today a major topic of debate and concern. In countries such as India, Bangladesh and Puerto Rico, for example, the rate at which the population is increasing exceeds all previous estimates and represents a problem of most alarming proportions. The application of modern science to nutrition, clinical medicine and the conservation of public health have so reduced the former "natural checks", which matched high birth rates with equally high death rates, that in many parts of the world governments are today "faced with the alternative of either a multiplication of population so rapid as to outstrip resources, or the acceptance of regulation . . . by the deliberate restriction of the number of children born". Unless drastic action of some sort is taken "every advance in the economy is absorbed by newcomers to the population, so that efforts to alleviate existing poverty, hunger, illiteracy, unemployment, over-crowding and homelessness are of no avail. The birth-rate must, therefore, be restricted to match the restricted death-rate." Again, it seems that methods of contraception practised in more sophisticated countries are, for the mass of the population, "too difficult, too expensive, or too unattractive to practise: in the social conditions in which millions live it is cheaper to bear, and perhaps to lose, a child than . . . prevent its conception by western means; and certainly it requires less effort of will. Sexual abstinence is similarly out of the question. Sterilisation — when there are already four or five children to the marriage — alone can meet the need: no more children, and married life as before."[21] As a consequence, in India, for example, voluntary sterilisation is today encouraged by propaganda, medical counselling and financial inducements and may even be made compulsory.

But before we go any further, let us pause to ask ourselves what attitude the Christian should take to this whole question.

[21] Cf. *Sterilisation: an ethical enquiry* (published by the Church Information Office, London, for the Board of Social responsibility for the then Church Assembly in 1962), p. 20.

So far as I know, there is no direct prohibition of sterilisation — or even, for that matter, of castration — in the Bible; but neither is there of an institution such as slavery. So we must approach this subject on the basis of general principles: namely, that man is a created being who does not belong to himself, and has not got an unlimited or unqualified right to dispose of himself as he wishes. On the contrary, "his right is limited by the laws of his Creator . . . and by the nature of his destiny".[22] It follows that he may not treat his body, or that of another, in any arbitrary way; and he may not mutilate it except where this is necessary. Thus the basic question is what constitutes "necessity" in this context. Traditionally, it has been taken to mean that a man may not mutilate or remove any part of his body "unless this is essential for the good of the whole, and unless no other, less drastic, measure is eligible". And although no one takes this so far as to apply it, for example, to a troublesome tooth which might be treated in some less drastic way than by extraction, the basic principle involved is recognised in our common law, where it governs consent as well as action; for a man may not lawfully consent to "excessive" injury (that is, injury more than "necessity" demands) being inflicted on him. But it is morally arguable that the doctrine of "necessity" should not be restricted to a purely personal or individual dimension, as though the need or benefit could, or should, be assessed exclusively in terms of the individual to be sterilised; for there is much in the New Testament which shows that morality must include a social dimension, and in this context that would involve the interests of the individual's marriage partner, family, society and even, perhaps, the whole body politic of which he is a member. Inevitably, then, this makes the moral decision much more complex, and "may well drive us to refuse the simplicity of absolute prohibition which a narrow, personal application of the original principles might seem to dictate". In other words, the term "necessity" might be regarded as extending to three inter-related criteria: the need of the state to limit an excessive population; the need of the family not to jeopardise its meagre

[22] Cf. *Human Sterilisation: Some Principles of Christian Ethics* (Published by the then Church Information Board for the then Church of England Moral Welfare Council in 1951), p. 3.

resources by sharing them with more claimants than they can satisfy; and the need of the parents for a married life which is not overwhelmed by the number of their children, and in which the wife, in particular, can be a real person. And these are, of course, the same arguments as are advanced to justify contraception.[23] So what, precisely, is the difference?

The primary difference is in the fact that sterilisation is, in practice, virtually irreversible for most of the people concerned. In the case of women the possibility of putting the original operation into reverse depends, basically, on the method that was used; and although in the case of men the theoretical chance of a successful operation to reverse vasectomy is reasonably good, it seems distinctly unlikely that more than a comparatively small number of patients of either sex would be able to find a doctor with the necessary time and skill. As a consequence there is an obvious tendency not to resort to sterilisation until the family is already of fairly ample proportions — and this, of itself, limits its effectiveness in controlling the birth rate. Even so, however, there is always the possibility that several children will die or the individual concerned will be widowed and want to marry again; in which case the fact that he (or she) had been sterilised might be a severe blow to the new spouse or even to the marriage.[24] This is one among several considerations which should certainly restrain any Government from resorting to compulsory sterilisation — except, possibly, in the most extreme and urgent circumstances — and which casts a distinct shadow of suspicion on the "voluntary" sterilisation of the poor under the impulse of financial inducements. It is also relevant to observe that Committee 5 in the Lambeth Conference of 1958 "agreed that any government policy of compulsory sterilisation as a means of population control is unacceptable to the Christian conscience, at least in the present state of knowledge and understanding".

But the second reason why sterilisation is today a matter of discussion and debate is for genetic reasons. In at least half of the American States, for example, sterilisation statutes are in

[23] Cf. *Sterilization: an ethical enquiry*. pp. 16–20.

[24] This is why, in the U.S.A., men who are to be sterilised sometimes deposit semen in a semen bank.

force. Under most of these the compulsory sterilisation of certain types of persons is either mandatory or permissive, while in a small minority the consent of the individual, his spouse or his guardian is required. In some of the States the application of the laws is limited to the inmates of certain designated institutions, while in others they also cover those who are at large. The list of persons who may (or must) be sterilised in every case includes the feeble-minded and usually the insane, while in some States it also extends to epileptics, criminals, moral degenerates and sexual perverts. Only in Georgia is the sterilisation of those suffering from hereditary physical disease provided for; and only in Nebraska is there an express provision that an operation for castration may be ordered by the court in the case of a man convicted for rape, incest or any crime against nature. In some cases, moreover, the grounds stated are simply "hereditary", while in others they are "hereditary and social". Some of the earlier statutes, which tended to confuse sterilisation for eugenic reasons with sterilisation as a punishment or for the protection of the public (for which it is, in any case, singularly ineffective), were declared to be unconstitutional; but the tendency of the more recent legislation has been to eliminate criminality. That sterilisation statutes are capable of being constitutional was decided in the Supreme Court in *Buck* v. *Bell*,[25] in the course of which Justice Holmes made his famous remark that "three generations of imbeciles are enough". Laws similar to those in the United States are also in force in the Canadian provinces of Alberta and British Columbia, while sterilisation laws have also been promulgated in Scandinavia, in some of the Swiss cantons and in Japan. In Denmark, moreover, provision has been made for the voluntary castration of sex offenders. But it was in Nazi Germany that both sterilisation and castration were brought into complete disrepute by reason of their widespread and compulsory use in the interest of a racialist ideology.[26]

In this country the legal position seems to be that "therapeutic sterilisation is lawful, contraceptive sterilisation is

[25] 274 U.S. 200.
[26] Cf. Glanville Williams, op. cit., pp. 84 f.

unlawful,[27] and the position of eugenic[28] sterilisation is doubtful".[29] The Brock Report,[30] for example, concluded that the eugenic sterilisation of mental defectives was legal, but not that of normal persons. The most recent trend of opinion among geneticists, however, is that sterilisation is likely to have only a minimal effect on the improvement of the race. Thus Professor Berry insists that "Sterilisation of carriers of dominantly-inherited conditions can reduce the incidence of the condition in question but cannot eliminate it, since most cases are due to new mutation"; that "sterilisation of homozygotes for deleterious recessive genes has a negligible effect on the bulk of the genes in the population"; and that "even if heterozygotes could be detected, it is not irrelevant to note that we all carry on average four or five deleterious recessive genes", so "a concerted attack on all these would be somewhat drastic".[31] Responsible opinion, it seems, "finds far less justification for sterilisation on eugenic grounds now than was found a generation ago . . . Even in cases of rhesus or other haematic incompatibility, indeed, specialist opinion seems to be hardening against a readiness to sterilise, since it would be mistaken to assume that the birth of a second affected infant will follow upon the birth of a first."[32] It is also relevant to observe that there has recently been considerable controversy about the sterilisation of a minor girl, on medical advice, with her parents' consent. But advice of this sort may sometimes be given on strictly genetic grounds and sometimes simply to prevent the birth of illegitimate babies by a minor who is subnormal and at moral risk.

Then, again, there is sterilisation for reasons of personal convenience: that is, as a simple method of contraception. Moral rigorists would argue that "mutilation may not be resorted to as a substitute for the exercise of will, enabled by grace", and that "while the practice of contraception requires, in theory, a deliberate decision of the will on every occasion, the acceptance of sterilisation (like the zealot's resort to castration) requires one such decision, "after which no further

[27] *Sed quaere.*

[28] Or "genetic" sterilisation — because of the parents' fear of having a defective child.

[29] St John-Stevas, op. cit., p. 163.

[30] Cmd. 4485, 1934.

[31] Op. cit., pp. 4 and 5.

[32] *Sterilization: an ethical enquiry*, pp. 34 and 35.

occasions for the exercise of the will in this regard may arise", and thus "no more such occasions for response to grace or for obedience to the will of God".[33] Others would argue that it represents little more than a particularly safe and easy method of birth control, which would mean much to a family living in extreme poverty, in circumstances in which other methods of contraception seemed unavailable or ineffective.

In sum, it seems wrong to conclude that non-therapeutic sterilisation should be unequivocally condemned in extreme circumstances — either for eugenic or genetic reasons or as a method of population control. In all cases, however, the Christian must insist that it involves a very grave personal decision — particularly, of course, in the favoured West. And this decision not only involves three different factors (whether the time has come when the couple should stop having any more children; whether sterilisation is the only appropriate means; and whether it should be the husband or wife who should be sterilised) but also, for the Christian, the fact that our bodies are "temples of the Holy Spirit", that our bodily functions are God-given, and that they should never be mutilated except for some truly adequate reason. But the medical adviser should certainly interpret the doctrine of "necessity" in a realistic way, in terms of the particular case in which his advice has been sought.

Finally, then, we come to the still more difficult and controversial subject of abortion, which has recently been such a focal point of heated debate. And here, again, we encounter the clash between extreme rigorism and extreme permissiveness, with a wide variety of closely-reasoned attitudes in between. The rigorist position is still, in the main, upheld by the Roman Catholic Church: namely, that it is never right deliberately to take the life of an embryo or foetus, even if this seems to be the only way to save the mother's life — although, by a species of casuistry which I find singularly unpersuasive, it is regarded as legitimate to perform a hysterectomy on a pregnant woman if there is some defect in her womb, in spite of the fact that this will inevitably destroy the foetus. The opposite extreme is often associated with "Women's Lib.": namely, that it is a woman who carries the foetus, and it should be within her

[33] Ibid., p. 22.

absolute discretion whether to continue or end her pregnancy — so she may go to a doctor and demand an abortion in exactly the same way in which she may go to a dentist and demand the extraction of a troublesome tooth. And the intermediate views all attempt, in different ways, to hold two conflicting principles in balance: namely, that of respect for the life of the unborn child and that of respect for the basic interests of the mother — and, perhaps, the family into which it is to be born.

But one can never pursue this subject very far before coming up against what is, in many ways, the basic problem: when, precisely, does the embryo or foetus become a "human being", with the full protection afforded by such categorical Biblical prohibitions as "the innocent and righteous slay thou not"[34] and with the potentiality for "eternal life"? Mr Oliver O'Donovan, in a thoughtful monograph, lists some seven different suggestions which have been made about this: first, conception, when the individual genotype is established of the child that *may* be; secondly, implantation, when the zygote becomes attached to its mother's womb; thirdly, the stage at which the embryo becomes a foetus which has assumed a recognisably human shape; fourthly, animation, about which such fantastic views were at one time held; fifthly, viability, when the foetus, if aborted, might conceivably be kept alive; sixthly, birth, when it becomes universally recognised as an infant, and foeticide would become infanticide; and seventhly, about one year later, when the infant reaches a stage of development comparable with that attained by most animals at birth. Many Christians today seem to imagine that the "moment of conception" theory has always been the orthodox view of the Church; so it is important to note that, even in the strictest circles — as Professor Dunstan insists — "protection was matched to development; higher with foetal growth, lower in the earlier embryonic stages". In the Aristotelian tradition of delayed "animation", for example, the decisive point was fixed at between twenty-five and forty days for a male foetus and fifty to eighty days for a female! Again, Gregory of Nyssa, writing at the end of the fourth century about belief in the Holy Spirit, remarks that "it would not be possible to style the

[34] Exodus 23:7.

unformed embryo a human being, but only a potential one — assuming that it is completed so as to come forth to human birth, while as long as it is in this unformed state it is something other than a human being".[35] The Irish Penitentials, moreover, graded the severity of their penances on the same basis: "for the destruction of the embryo of a child in the mother's womb", three years on bread and water, and "for the destruction of flesh and spirit [i.e. the animated foetus] in the womb", fourteen years on bread and water.[36] And Dunstan adds that "the distinction between *foetus animatus* and *foetus inanimatus* or *informis* persisted unbroken in Roman Catholic tradition until the decrees of 1884 to 1902 — except for three years between the issue of an 'absolutist' decree by Sixtus V in 1588 and its revocation by Gregory XIV in 1591. Absolute protection for the foetus at every stage has not been guaranteed for any but a very short time in the Christian moral tradition."[37] So we must, I think, conclude that — however much O'Donovan objects to an attitude of "agnosticism" on this matter — it is in fact impossible to fix any particular time when the embryo or foetus "becomes a human being", before which abortion is of little or no consequence and after which it becomes murder. All we can safely assert is that it is, right throughout pregnancy, at least a *potential* human being, and should therefore be treated with all due respect. So it would follow that where there is a *genuine* need to choose between the life of the foetus and grave danger to the life (or fundamental well-being) of the mother, it must surely be right to give priority to the mother, as a fully developed human being. But to this basic consideration — and its repercussions — we must return later.

The problem that concerns me at the moment, however, is whether we cannot perhaps go a little further than this in examining the process by which the embryo "becomes a human being". To begin with, as Dunstan puts it, it is "hardly

[35] *Adversus Macedonianos*, trans. from Library of Nicene and Post-Nicene Fathers, Ser. 2, vol. V, ed. H. Wace and P. Schaff (Oxford and New York, 1893), p. 320.

[36] *The Irish Penitentials*, ed. Ludwig Bieler (Scriptures Latini Hiberniae, Dublin, 1963), p. 229.

[37] Cf. op. cit., pp. 82 and 83.

apt to speak of a 'moment' of conception. Conception is rather a process, beginning with the quick passage of the sperm into the cytoplasm of the ovum and extending surely over the first week of cell division". In any case, he insists,

> popular language about "soul" and "body" as two entities is not traditional Christian language at all; it came into the Western tradition through Platonism from remote primitive sources. So long as the Aristotelian tradition dominated Christian thinking, the notion of "immediate animation" would have been repudiated: animation, too, was a process, and a lengthy one. For the Aristotelians, *psyche*, *anima*, meant, not an entity called "the soul", but the principle of organisation, that which gave a "thing" and its activities their characteristic form, "the first principle of life in living things about us". "The range of activities of a living thing reveals the kind of soul which is present within it", writes F. Copleston, interpreting St Thomas Aquinas . . . The rational soul is what declares the body to be a human body, and soul and body are one substance.

So much for the Aristotelian tradition which dominated Christian thinking for so long. But modern embryologists, Dunstan continues, commonly speak of a human embryo becoming a foetus "when it acquires a recognisable human form", about 3 cm long, at about forty-five to forty-nine days. Undoubtedly the full *potential* for human personality is present chromosomally from the time of fertilisation; but we "do not know at what point in the nine months in the womb the conceptus becomes 'human'. We have to reckon, anyway, with the estimate that 30%, perhaps up to 50%, of zygotes, fully fertilised human cells, are lost before implantation."[38]

What, then, would this seem to imply in regard to two matters which have already concerned us: the experiments in regard to the fertilisation and growth of a zygote *in vitro*, to which reference was made in my last lecture, and the destruction of zygotes in the womb at a very early stage, which I have already discussed briefly in this? We can certainly postulate that human life is uniquely precious to God, and that no one

[38] Dunstan, op. cit., pp. 68 and 69.

should procreate life irresponsibly, whether by natural processes or in a laboratory. But it is, I think, important to remember that scientific experiments with sperms and ova are, in most cases, not conducted out of any desire to "play God", but rather to discover biological secrets which may be of great interest and value. They should, indeed, be pursued in the spirit of Newton's wonder at the privilege of thinking God's thoughts after him — although we may well take a different view of some of the purposes to which these experiments might subsequently be put. So it seems to me distinctly doubtful if we should condemn such experiments as necessarily demonstrating any disrespect for human life just because ova may be fertilised *in vitro*, kept alive for a few days, and eventually allowed to die. In wrestling with the problems raised by such experiments Dunstan discusses the question whether there is a point where the legitimate search for knowledge "must cease because it begins to violate a human claim" — a point when "to maintain the life of a human organism outside the womb would become morally intolerable". So he considers the further question whether there is a point

> at which the foetus becomes relationally dependent on its mother — dependent on her, that is, for more than a chemical environment, oxygen, nutrient fluids, hormones and the like, dependent on her as a person. When does her *presence* to the foetus begin to awaken in the foetus the potential for human response, as awareness of the foetus begins to awaken in the mother the beginning of a *maternal* response? If we could know at what point a mother, as a human being, as a source of specifically human relationship, becomes irreplaceably necessary to the development of the human embryo into a human child, then we should see a threshold at which experiment must cease, a step which must not be crossed. For beyond it lies the life of a man, the image and glory of God: and this is holy ground.[39]

When we apply this principle of respect for life to those cases of contraception which virtually — or actually — represent an exceedingly early abortion, somewhat the same

[39] Op. cit., p. 71.

principles would seem to apply. We have no right to procreate life irresponsibly — whether inside or outside marriage. But the prevention of a zygote from implantation, or from the very early stages of development, seems to be very different from the deliberate destruction of a foetus which has, in Karl Rahner's words, become "a being with all the basic rights of a human person" — a status which, in his view, cannot with any certainty be accorded to an embryo during the first month or so of its development.[40] Clearly, however, the gravity of any such intervention increases with every week of pregnancy, for we now know that an embryo has a distinct heart beat at seventeen days and that its brain produces a measurable electro-encephalogram at five weeks.[41]

Now it has already been remarked in passing that a convincing case for abortion can be made out, in my view, when the life or fundamental health of the mother is at stake — on the simple principle of the priority which should be accorded to a developed human personality *vis-à-vis* one in course of development. But the relevant section in the Abortion Act of 1967 allows abortion where two registered medical practitioners believe, in good faith, "that the continuance of the pregnancy would involve risk to the life of the pregnant woman, or of injury to the physical or mental health of the pregnant woman or of any existing children of her family, greater than if the pregnancy were terminated". This wording seems to me totally inadequate, for two principal reasons. The first is that the criterion of striking a balance between the alleged risks involved in terminating the pregnancy, on the one hand, and letting it go full term, on the other, is so vague as in practice to open the door wide — contrary to the stated purpose of the Act — to abortion on demand. It provides no real test for a genuinely therapeutic abortion — based on some serious danger to the mother's life or health. And the second weakness in the section is the reference to injury to the physical or mental health of "any existing children of her family", for this, again, is a criterion wide enough to allow a coach and four to be driven through it. It seems to me clear that this clause

[40] Cf. Häring, op. cit., pp. 75 ff.

[41] Cf. Harley S. Smyth, in *Biblical Allusions to Life before Birth* (published by the Christian Medical Fellowship, London, 1975), p. 6.

should be eliminated, and that the interests of the "existing children of the marriage" should affect the decision only in so far as the apparent impossibility of coping with her family problems has, in sober fact, put the pregnant woman's physical or mental health at genuine risk — a criterion which could, and in my view should, also be applied to the pregnancy of one who is very young, or a pregnancy which is allegedly the result of rape, etc. After all, a doctor's competence can scarcely be said to extend beyond the health — physical or mental — of his patient.

A further section in the Act, however, provides that abortion is also justified where two registered medical practitioners believe, in good faith, "that there is a substantial risk that if the child were born it would suffer from such physical or mental abnormalities as to be seriously handicapped" — and this raises the whole question of abortion as part of the negative approach to "genetic engineering".[42] Put in its simplest terms it may be said that "more than forty inherited diseases (metabolic or chromosomal) can now be identified by culturing cells drawn from the amniotic cavity of pregnant women".[43] It seems, however, that amniocentesis, or the process of obtaining a sample of amniotic fluid from around the foetus, is not feasible before the fourteenth week of pregnancy and that most tests are carried out at sixteen weeks — after which a further period is required for diagnosis.[44] The process is not, moreover, wholly without danger; although this has now been reduced by an ultrasonic technique by means of which the position of the foetus can be determined before the instrument is inserted. In any case, amniocentesis is normally used only in the case of women who, for one reason or another, are thought to be particularly at risk — either because of some chromosomal or genetic defect in one or both of the parents, because the woman concerned has already given birth to a defective child, or because she is over forty (since the danger of such a woman bearing a mongol child is twenty times greater than in the case of a woman of twenty-five).

[42] Including in this term action taken on behalf of an individual child rather than the race as a whole.

[43] R. J. Berry, op. cit., p. 5.

[44] *Our Future Inheritance*, p. 59.

But the ethical problems inherent in abortion by reason of some actual or potential defect in the foetus seem to me particularly difficult. Mr D. M. Jackson, for example, states that he would not himself feel that abortion was "justified by correctable foetal deformities or deformities compatible with the enjoyment of life. I have several children under my care with arms and legs burned off, but in spite of their handicaps they are delightful little people, very lovable and much loved . . ."[45] Even under the Act, moreover, the risk has to be "substantial" and the handicap to be "serious" — so there is the problem of what may be termed the *statistical* risk. Apart from those chromosomal disorders or enzyme deficiencies that can be positively diagnosed by amniocentesis (and this is now possible even in cases of spina bifida), and those (like gross hydrocephaly) which can be diagnosed only very late in pregnancy, Dunstan states, "the most that can be predicted is a statistical risk of defect — as high as fifty per cent for the rare conditions, like haemophilia (if the foetus is male), but much lower for other disorders, like those arising from maternal contact with rubella. To terminate on the basis of statistical risk is to accept the necessity of killing more healthy foetuses than defective ones."[46] An additional hazard is that in cases of rubella (or German measles) the morbidity rate in terms of the foetus varies enormously from epidemic to epidemic. But we must remember, in all cases in which a foetus is, or may be, "seriously handicapped", that two different issues are usually at stake: the handicap which the child is likely to suffer, and the effect of this handicap on the mother and her whole family — which may, in some cases, be catastrophic. And here, as in so many cases of possible abortion, two conflicting principles may make the problem an agonising one. On the one side there is the deep compassion we must always feel for a woman driven almost to despair by a totally unwanted pregnancy or by the daunting prospect of caring for a helpless child; and on the other there is the principle that hard cases make bad law — and, indeed, that it is virtually impossible to draft legislation which would afford a conscientious doctor

[45] *Human Life and Worth* (Christian Medical Fellowship, London, 1958), p. 14.

[46] Op. cit., p. 85.

adequate scope for a responsible exercise of his discretion without leaving loopholes for possible evasion and abuse. If reverence for human life must prevent us from putting an imbecile to death, or deliberately sacrificing the life of a defective baby which has already been born — and points such as these will engage our attention in my next lecture — then are not much the same principles applicable to the destruction of a foetus which is already in an advanced stage of development?

Four other aspects of this problem must also be mentioned in passing. First, I have already stated that genetic experimentation in breeding animals was readily acceptable — provided this involved no cruel exploitation — because any unfortunate result could be remedied by the simple expedient of putting the animal concerned painlessly to death. But this would clearly be impossible in the case of human beings — so the alternative sometimes suggested is widespread abortion. But surely this is almost equally unacceptable from an ethical point of view? Irresponsible experimentation, with abortion as the readily available remedy for mistakes, is totally unworthy of man made in the image of God. Secondly, there is the question of the opportunity provided by abortion for research on foetuses — about which there has been much discussion. From the ethical point of view, however, the issue in this case seems perfectly clear. It would be manifestly wrong, in my view, for a doctor to perform or advise an abortion in order to facilitate such research; but there could equally be no moral objection to a foetus, legitimately aborted, being used — like any other dead body — for responsible research. Thirdly, it seems to me that the decision in regard to an abortion must — normally at least — be left to the mother and the appropriate medical authorities rather than to a committee; that no mother should be pressurised into having an abortion against her will; and that no doctor should be compelled to perform an abortion against his conscience or judgment — or be liable to legal action in regard to a decision he has conscientiously made. Nor should a doctor be penalised because of his views in this matter — except that a consultant's rigid refusal to perform any abortions whatever might be a factor which must inevitably be taken into account in his appointment to a position in which he was virtually in sole charge. Fourthly, it is imperative that we should appreciate the impact that a permissive policy in regard to abortion must necessarily have on the value that society attributes to human life as such. But I shall have to revert to this whole subject at the beginning of my next lecture.

Clearly, then, whatever the law may decree, the question of abortion poses an exceedingly complex ethical problem for mothers, doctors and society at large. At the very best, an abortion can never be regarded as other than the lesser of two evils — an agonising problem thrown up by a very imperfect world. The basic consideration for the Christian must always be the will of God, but this does not relieve us from coming to a responsible decision — for we should not leave everything to providence, in a fatalistic way, when God has given us the knowledge and ability to take appropriate action. What, then, should we do in any particular case? The only answer I can give at this point is that we must grapple with the problem before God earnestly and responsibly — but with the scales weighted, I think, always in favour of reverence for human life.

4 The Prolongation of Life, Transplant Surgery, Euthanasia and Suicide

My last lecture was concerned with the prevention or control of the inception of human life by means of contraception, sterilisation or abortion. But it is clear that abortion, in any form, also represents the termination of life — although the question of whether, and precisely when, the life of an embryo or foetus should be regarded as one which is fully human is, as we have seen, a matter of much uncertainty and debate. There can be no doubt, however, that there is an inescapable connection between abortion and the whole problem of the prolongation or termination of life by medical intervention — not only, as Dunstan insists, because abortion and euthanasia "both presuppose the same right over life", but also because today "both are demanded in terms of 'liberalisation' of the law". I believe he makes a most valuable contribution to the whole subject of this lecture, moreover, when he suggests that the ethical problems involved can be more meaningfully discussed if we think of abortion as "foeticide" and consider it in the context not only of homicide and suicide, but of "infanticide", in regard to the killing of infants born with some gross congenital handicap, "senilicide", in regard to the killing of the very old, and some such word as "dementicide" or "amenticide", in regard to the killing of those who may be judged to be mentally unfit to live.[1] The difference, some would hold, is merely one of degree, while others would see it as the difference between killing a person and a "non-person" — wherever, precisely, they would draw the line.

Now I might well have concluded my last lecture by venturing to suggest:

(1) That we can scarcely consider those methods of contraception which operate by preventing the zygote, or fertilised ovum, from becoming attached to the wall of the womb — or even (if research to this end if successful) by means of a pill taken once a month to ensure that, should implantation have taken place, the zygote is almost immediately removed — as the destruction of "human" life in any very meaningful sense of that term, whatever view we may take of the relative desirability or undesirability of these methods.

(2) That in the considerable period between the earliest days of pregnancy and the time when a foetus becomes viable, abortion becomes increasingly abhorrent, as the foetus grows progressively more human; and that it should be allowed only where it seems to be essential, on strictly therapeutic grounds, because of some genuine danger to the life or health of the mother, or conceivably on reliable evidence of some really grave malformation in the foetus itself — where the mother and appropriate medical authorities are agreed.

(3) That once the foetus has become viable, "feticide" becomes scarcely distinguishable from infanticide, and should be allowed only in the most extreme cases — e.g. where there is no other way in which the mother's life or absolutely basic health can be preserved, even by a Caesarian operation, or where the foetus is so grossly malformed that, if born, it could not live or could scarcely be regarded as human.

So our starting point in this lecture must, I think, be the attitude which we should take to a baby which has been born alive but with some serious physical or mental handicap. And here we must attempt to distinguish three different categories of cases and circumstances: first, those in which it would be right and proper for no steps whatever to be taken to preserve his life, but for him to be allowed to die virtually before he has begun to live; secondly, those in which his life should be

[1] Op. cit., p. 88.

preserved by all normal forms of care but, should some infection or other hazard to his survival subsequently intervene, it would be bad medicine to resort to antibiotics, surgery, or other special medication (the proper use of which is to assist a patient through a critical period towards a recovery of health — or, at least, a meaningful extension of life), in order, in this case, only to prolong what is virtually the process of dying; and, thirdly, those in which every effort should be made to preserve his life, whether by surgery, diet or other treatment designed to remove or alleviate his handicap — or at least to ensure that even a severely handicapped life is made as tolerable and useful as possible.

If, moreover, we can make these distinctions in a way that is even tentatively acceptable, this will clearly throw some light backwards, as it were, on whether the attitude we have taken in regard to a deformed or genetically defective foetus is right, and possibly also forward, *mutatis mutandis*, on the problem of what should be done in the case of those who are no longer really alive in any meaningful sense of that term, those who are suffering from a terminal illness whose lives can only be somewhat prolonged by very special medical or surgical intervention, or those who are dying from extreme old age.

First, then, in what circumstances, if any, should a newly born infant be allowed to die virtually before he has really begun to live? To this the answer which is usually given is where the baby can be classified as a "monster" (that is, one born with no proper brain or so grossly deformed as not to be human in any meaningful sense of that term) or where it is so malformed that it could not possibly survive for more than a minimal period — for in such cases, I understand, no action is normally taken either to stimulate life or to preserve it (except that, if the baby shows an active desire for food, this is, I believe, usually provided).

Now it is obviously possible to argue that no valid distinction can be made, in moral theory, between killing a baby (or any other creature) by some positive action, on the one hand, and deliberately allowing it to die, by withholding some necessary action, on the other — although the law regards the actual killing of a newly born baby as murder (or, if done by the mother, as infanticide — in a technical use of that term), while it does not regard a failure to intervene to save anyone's life as criminal except in so far as the one who fails to act is under a duty of care to the person concerned. And while a doctor, midwife or nurse clearly has a duty of care in regard to a baby in their charge, yet should the baby concerned have been born with virtually no brain, or so grossly deformed as to be incapable of carrying out normal functions, that duty would, presumably, be properly fulfilled by such action — or inaction — as was medically appropriate in these very exceptional circumstances.

But what light, if any, do such cases throw backwards on the problem of what should be done in regard to a seriously defective foetus, and forwards on the problem of the action that should be taken in regard to the very old, the very ill or those who have lapsed into a coma from which they will never recover? In the case of a foetus which can be conclusively proved — whether by amniocentesis or some other means — to be so radically malformed that, should it come to birth, it would be allowed to die at once, an abortion would presumably be regarded as permissible by everyone except such as may believe that even a few moments of independent life — by contrast with any period of foetal life — constitute an essential prelude to life beyond the grave. From any other viewpoint, the decision in such cases must surely depend entirely on what the mother desires and her doctors deem to be in her best interests, both from the physical and psychological point of view. But the only situation in regard to terminal patients comparable to this is where the person concerned can no longer be regarded as still alive in any meaningful sense; for the basic requirements of food, warmth and cleanliness should never be withheld from such persons before death has actually intervened. We shall, however, have to return to this subject in much more detail later, since it involves the complex problem of how, precisely, death is to be defined and determined today.

Secondly, there are a number of different circumstances in which a baby's life should certainly be preserved by every normal form of care, but where it is probable that no attempt should be made to prolong a very defective and precarious life by surgery, antibiotics or similar forms of medical intervention which are designed, as we have seen, to help a patient through a period of acute sickness. It is clearly beyond the scope of this lecture — or, indeed, the competence of the lecturer — to go into details, and two or three examples must suffice. In a disease called spina bifida, for instance, the spinal cord and nerves are so exposed that in the worst cases death commonly occurs at a very early age. In a significant number of cases, however, these outward defects can be corrected by an operation, but only at the risk of augmenting brain damage which leaves the child gravely retarded.[2] A children's hospital in Sheffield has specialised in the treatment of this disease over a considerable period, and for ten years its policy was to accept every child which was brought to it and always do everything possible to lessen its handicaps. But the result, in some cases, was to prolong and intensify suffering into the years of greater awareness and sensitivity, since each intervention created the need for yet another at a later date. So the policy of maximum treatment for all has now been abandoned in favour of one of selection, based on the clinical prognosis of the degree of anticipated handicap. But this does not mean that some children are left to suffer untreated, or still less that their life is deliberately brought to an end.[3] The so-called "contra-indications to active therapy" were officially accepted in 1973 by the Department of Health and Social Security as an allowed practice, and are based on the principle that the extent of the damage to the child at birth should be carefully assessed and then, in the light of experience of how such children have fared in the past, a line should be drawn above which every effort should be made to save the life of the child concerned, and below which no such action should be taken. As a result those who are treated have a reasonable chance of living a life which is not too unpleasant, while those who are not treated generally die within a year of birth — although in a few cases they have survived for considerably longer. In Sheffield, moreover, the parents of affected children are given a full explanation of the situation, including the possible therapeutic treatment; a prognosis of the likely minimum handicap that their child might have if treatment is offered and if everything goes according to

[2] *Our Future Inheritance*, p. 58.
[3] Cf. Dunstan, op. cit., p. 90.

plan; and a recommendation by the doctor, but with the option of requesting a second opinion.[4]

Here two basic principles, it seems to me, are involved, both of which are of wide application and must later be discussed in more detail. The first is how limited medical resources can best be apportioned; for it is obvious that treatment which is very costly, both in money and manpower, cannot be provided on a vast scale. It is a matter of the utmost ethical importance, however, to determine the criteria on which the selection of prospective patients should be made. And the second principle concerns the duration and quality of life which is likely to result from painful and, perhaps, repeated or prolonged medical intervention. This, too, gives rise to a problem of wide application, which is especially relevant to terminal illness; for in such cases doctors are often faced with the need to decide whether the stage may not have been reached when all concerned should concentrate on the symptoms rather than the disease — and should direct their care to allowing death to supervene with as little pain or distress as possible, rather than attempting to ward it off a little longer by means which would only cause unnecessary suffering.

But to return to infants. As medical care improves it becomes increasingly possible to keep deformed and damaged infants alive for indefinite periods; and it has been reported that even an anencephalic child — that is, the extreme case (to which reference has already been made) of a baby born with virtually no brain — has been kept alive for several weeks. It is now possible, moreover, for doctors to detect most cases of spina bifida and anencephaly *in utero* by amniocentesis; but it seems clear that the cost of amniocentesis, and all that it involves, will preclude any routine tests being made available to pregnant women except in those cases in which some method has been found for identifying mothers at risk. It may, however, soon be possible to detect both these diseases much more easily and cheaply, if with somewhat less certainty, by simple testing the blood of the expectant mother, and this would mean that cases of anencephaly and spina bifida could be predicted as a matter of routine in plenty of time for the foetus to be aborted.

[4] *Our Future Inheritance*, pp. 123 f.

In regard to anencephaly there could, presumably, be no moral objection to abortion if the mother so desired; but cases of spina bifida seem to fall into a different category — partly because the child concerned is in no sense a "monster", and partly because it would often be exceedingly difficult to detect, *in utero*, how severely the child, if born, would prove to be afflicted. Cases of mongolism can also be detected *in utero* by amniocentesis; and this is now normally available to women who are known to be at risk — provided, in most cases, that they agree in principle that they are prepared to undergo an abortion if the prognosis is unfavourable. It is obviously relevant to note, in this context, that the cost of providing life-time institutional care (where this is in fact necessary) to a mongol is said to amount to something in the order of £100.000; but mongols are frequently very affectionate, arouse a corresponding love in their parents, and exercise a humanising influence on those who care for them.[5] To let a mongol die after birth would clearly amount to murder or infanticide; so it would be exceedingly difficult, I think, to find any moral justification for abortion in such cases — unless the mother-to-be, in her particular circumstances, felt so utterly unable to cope with the anticipated strain that her mental health was in genuine jeopardy; and this, of course, would mean that the abortion would be on therapeutic, rather than on purely genetic grounds.

Cases of cystic fibrosis of the pancreas (which causes the obstruction of various ducts and organs by mucus) and phenylketonuria or "PKU" (a disease caused by the body's inability to metabolise the amino acid phenalalanine), on the other hand, cannot as yet be detected in the womb. Both can, however, be diagnosed, in principle, by a standard test after birth; and in this country, theoretically at least, all babies are today tested for PKU during the first few weeks of life. If cystic fibrosis is detected at a very early age it is now possible to give medical or surgical treatment by which the life-expectancy of sufferers has been progressively extended; but very few survive the age of about twenty. Providing PKU can be detected soon enough, on the other hand, the prognosis is much more favourable; for while, in the more serious cases, a child born

[5] Cf. Dunstan, op. cit., p. 85.

with this genetic defect will, if left untreated, always become severely mentally retarded, this can be counteracted by a special diet.

To sum up, then. There are a few cases in which babies are born which are so radically defective, mis-shapen, or "monstrous" that the only proper action seems to be to allow them to die virtually before life has begun. In other cases of very grave handicap or defect, on the other hand, they should be given every normal form of care, love and attention; but it would seem contrary to the principles of compassionate medicine to take any extreme action to prolong life should some crisis of health intervene. In yet other cases of even severe handicap, however, every effort should be made to enable the sufferer to develop his existing faculties to the full — whatever the cost. Here Helen Keller is, I suppose the classical example; for after losing sight and hearing at the age of nineteen months, allegedly through scarlet fever, she learnt to read, to write and later to speak; she got a university degree at the age of twenty-four; and she subsequently became a lecturer of international fame who has been an inspiration to thousands. Surely, then, much the same principles are applicable, *mutatis mutandis*, to questions of genetic abortion. The basic principle must always be the preservation of at least a potential human life; and this principle should be displaced only where the circumstances are such that the life or *essential* health of the mother is genuinely at stake, or the malformation of the foetus is both certain and extreme. It is precisely at this point, moreover, that medical research should be so warmly supported; for such research makes it progressively more and more possible to safeguard the life and health of the mother and to establish with certainty whether a foetus is in fact gravely malformed rather than potentially at risk — whether because the mother has been exposed to rubella, because one or both parents are known to carry a recessive defective gene, or because of a number of other possible factors. Even so, abortion will, no doubt, continue to be recommended in practice in many cases in which no ethical justification exists; but there will at least be less and less excuse for the wholesale destruction of perfectly healthy foetuses in order to ensure that a comparatively minor number of malformed children are not born. It is obvious, moreover, that the considerations which can be adduced to support genetic, as distinct from therapeutic, abortion could also be used to justify genetic infanticide, unless a clear distinction can be made between an embryo or foetus in the early, middle or late stages of pregnancy, respectively, and a child that has actually been delivered.

Somewhat the same arguments, moreover, could also be adduced to support what Dunstan terms "senilicide" and "dementicide" — or, indeed, voluntary euthanasia. Strictly speaking, of course, euthanasaia simply means a quiet and easy death — or, in terms of a valuable report produced in 1975 by a working party set up by the Board for Social Responsibility of the General Synod of the Church of England — "dying well". But in practice the word is normally used today of the means of procuring, or the action of inducing, such a death. If applied to infants, the insane, or those in coma, the qualifying adjective "voluntary" would, of course, be wholly out of place; and there is always the possibility that a future Hitler or other ruthless pragmatist might decree that all those judged to be worthless or burdensome to the State should be painlessly eliminated. But for all practical purposes the current debate centres around the demand that "voluntary", as distinct from any form of compulsory, euthanasia should be permitted by law — although as we shall see later, this might well be the thin edge of the wedge.

Now at first sight a persuasive case can admittedly be made out — except, perhaps, from a specifically Christian viewpoint — for removing any criminal sanction from voluntary euthanasia. The literature of the Voluntary Euthanasia Society abounds with moving stories of those whose lives have become increasingly burdensome both to themselves and others, or even virtually intolerable by reason of constant pain in circumstances in which there is no conceivable prospect of recovery, who beg their doctors to give them a lethal dose — followed by a description of the relief of all concerned when he complies, and the sense of guilt he experiences when he feels compelled to refuse. Other stories describe how patients are sometimes subjected to painful surgery, or distressing medication, which only serve to prolong, for a very limited period, a life which is not really worth living, instead of being allowed to die in peace; of elderly persons whose arrested hearts are reactivated by massage in circumstances which do little more than give them the doubtful privilege of "dying twice"; of those who can never recover consciousness but whose biological life is maintained by some artificial means; and of people so distraught that they deliberately attempt suicide only to be forcibly resuscitated and thereby driven to make the same attempt again in even more distressing circumstances. Cases like this, it is said, may provide outstanding examples of the way in which doctors fulfil their duty to preserve and maintain life but fall very far short of fulfilling their complementary duty of compassionate care for the suffering. Now that suicide and attempted suicide are no longer regarded as criminal, moreover, the refusal to regard a properly regulated right to request euthanasia in the same light may appear somewhat anomalous.

The only satisfactory way to approach this difficult subject is to begin by making a number of essential distinctions — somewhat along the same lines as those which we have already discussed in the context of malformed or defective babies — and then to examine the viable alternatives in the light of their medical, legal, social and ethical implications. And the first distinction we must make is between euthanasia in the proper sense of the current use of that term (that is, the deliberate decision to end a patient's life by what is sometimes called "mercy killing"), on the one hand, and a wide variety of other forms of treatment which may allow death to supervene, on the other. It is widely recognised in the medical profession today, for example, that a surgeon contemplating a palliative operation on an elderly patient should ask himself the question "Would this procedure give him a *reasonable* chance of an *appreciable* duration of *desirable* life at an *acceptable* cost of suffering?"[6]; for although none of these adjectives can be exactly defined, all of them should be taken into consideration in the course of making a responsible medical decision. In so

[6] Cf. D. M. Jackson, op. cit., p. 10. On this whole subject of "meddlesome medicine" — and, indeed, the many and complex problems raised by euthanasia in all its aspects — cf. Hugh Trowell, *The Unfinished Debate on Euthanasia* (SCM Press, London, 1973).

far as this is possible, moreover, the whole position should be explained to the patient, who always has the right, when conscious, to refuse to accept his doctor's advice. Similarly, when the life of a patient suffering from terminal cancer is threatened by an attack of pneumonia, for example, it is wrong to describe as "euthanasia" a doctor's responsible decision that his duty of care is best fulfilled by withholding antibiotics — or, in other circumstances, that the time has come to cease to grapple with the fatal disease and to concentrate exclusively on its symptoms, and thus to allow the patient's life to draw to a peaceful close. And precisely the same principle applies to the administration of whatever drugs, in whatever quantities, may be necessary to relieve severe pain. In many cases these drugs may, of course, have the side-effect of marginally shortening the patient's span of life, although sometimes the relief of pain and anxiety have precisely the opposite effect; but there is, in any case, an essential difference between a determination to relieve suffering in order to minimise the trauma of death and a deliberate decision to precipitate death in order to end the trauma of suffering.

Now it is, of course, perfectly possible — as we have already noted in passing — to argue that it is exceedingly difficult to make any valid ethical distinction between causing a person's death by some positive act, on the one hand, and allowing him to die by some deliberate refusal or failure to act, on the other. But this argument is, I believe, wholly out of place in the present context, where allusions to what is sometimes called "passive" or "indirect" euthanasia are most misleading. The dilemma with which the doctor has to grapple is not an abstract problem in the field of ethical theory but the practical question of what is the proper treatment for an individual patient in particular circumstances. And here a number of different factors are involved. He has a clear duty, for example, to preserve his patient's life in any acceptable way, but this does not imply a duty to prolong the process of dying by what may aptly be termed "meddlesome medicine". Again, he has a complementary duty to relieve his suffering by any legitimate means, but these do not include the deliberate prescription of a lethal dose.

But why, it may be asked, should we make this solitary

exception, when circumstances undeniably exist in which this would seem to many people to be the obvious solution? A pertinent answer to this question is that the overwhelming majority of doctors would, I think, strenuously oppose the introduction of any legislation designed to authorise voluntary euthanasia — for a number of different reasons. First, there would be the problem of when and how a patient should make his request. If the law were to authorise a doctor to give a lethal dose on demand, or after a brief interval, at a time when his patient happened to be in special pain or despondency, there is a wealth of experience to show that the patient would in many cases have taken a different view a few days later; but it would then be too late. If, on the other hand, the patient were required to sign a form stating that, should he ever be afflicted with some incurable and painful disease or crippling disability, he wished his doctor to exercise his discretion as to when he should put him out of his misery, then a number of other problems would arise. How, for example, is a doctor to be sure that his patient's disease is incurable, or whether the degree of pain that he is suffering is sufficient to justify such an irreversible intervention? And what effect is such a request for future action at the doctor's discretion likely to have on the trust and confidence which are such vital factors in the relationship between patient and doctor? Would not most people who had signed such a request begin to wonder, as soon as they felt sufficiently ill, whether every pill, dose or injection that the doctor prescribed was designed to cure or kill, to relieve their pain or bring it to a final end? There is, again, plenty of experience to prove that even patients who firmly believe in voluntary euthanasia, and say they hope it will be available for them one day, commonly refuse any offer that the fatal dose should immediately be administered. Even Glanville Williams' suggestion that, where a patient is "suffering severe distress without prospect of relief . . . it should not be murder for a doctor, acting at the patient's request and with the concurrence of another doctor, to accelerate the patient's death in good faith and for his benefit", is open to a number of objections, including the virtual impossibility of defining the term "no prospect of relief" and deciding where and when it is applicable.

Then, again, there is the problem of the circumstances in which any such request should really be regarded as "voluntary". Would not an aged parent whose care seemed to impose an almost intolerable burden on a daughter, for example, often feel under considerable moral pressure to relieve her of that burden were it lawful to ask a doctor to intervene in this way? But is it right that any such dilemma should be added to an experience which is already, in all conscience, sufficiently harrowing? And might not the person who has to bear the burden sometimes be sorely tempted to feel that the parent ought to be sufficiently unselfish to make the request? It would be all too easy for a passing feeling of this sort to betray itself in some word or look — or, at least, for the parent to interpret some gesture or expression of impatience, disgust, frustration or sheer weariness as an indication of some such feeling. It is precisely in circumstances such as these that the law helps to formulate, sustain and re-inforce morality.

It may, of course, be argued that it would often be perfectly possible for the patient, in such cases, himself to take a lethal dose without involving the intervention of anyone else. But comparatively few people have the means or the knowledge to enable them to do this; and many of those who could would hesitate actually to commit suicide — whether for reasons of personal conscience or to spare the feelings of those they love. There is a real, if subtle, difference between the situation which now prevails, in spite of the fact that suicide is no longer a crime, and the climate of opinion which might come to prevail were legal provision to be made for voluntary euthanasia. Even from the strictly legal point of view, moreover, it is a fallacy to argue that the law now recognises a "right to die" comparable with the "right to live"; for a person who "aids, abets, counsels or procures" the suicide of another, or even an attempt to commit suicide, is liable under the Suicide Act, 1961, to a term of imprisonment of up to fourteen years, while it is still murder actually to kill another person even at his urgent and insistent request.

The Christian, in particular, has always regarded suicide as morally wrong, and there can be little or no ethical difference between suicide and voluntary euthanasia. It was only a few years ago, indeed, that suicide was widely regarded as a form of murder and attempted suicide as an attempt to commit murder — although it was judicially decided in 1862[7] that attempted suicide was not to be so regarded within the meaning of the Offences Against the Person Act, 1861. As a result, attempted suicide came to be classified as a misdemeanour rather than a felony, with the happy result that no one was under a legal obligation to report such an attempt to the police. All the same, one who attempted to commit suicide was liable to prosecution and imprisonment; and although the successful suicide had clearly placed himself outside the reach of the criminal law, in the past his estate was forfeited, he was denied the rites of Christian burial and his corpse was interred at cross-roads with a stake driven through his heart! This was partly because suicide was regarded as necessarily constituting an "unrepented" sin, and partly as a relic of pagan superstition.

Happily, a much more sympathetic and understanding attitude towards those who resort to suicide, or attempt it, in a state of deep mental distress, depression or despair has now come to prevail; and this has been reflected in the Suicide Act, which recognises that such people stand in need of succour (medical, social and spiritual) rather than punishment. In many cases, indeed, "attempted suicide" represents, in reality, a clamant cry for help. But this change of attitude should not be taken to give any support to the claim that men and women have a right to do what they like with their lives and that suicide never represents a sin or incurs moral blame. In point of fact the taking of one's own life can range all the way from an act of supreme self-sacrifice to a form of cowardly and selfish escapism. We all admire the nobility of Captain Oates, for example, but it is difficult not to take a very different attitude to what has been termed "the coward's way out" — although our instinct to judge and condemn is softened and mitigated by a realisation that it is virtually impossible to put oneself inside the mind of another person or to appreciate the stress, pressure and anguish which may have seemed to drive him up to, or over, the brink. It is insensitive legalism to speak of suicide — or, indeed, any genuine instance of "mercy killing" — as "murder" when viewed from the perspective of

[7] *R.* v. *Burgess* (L. and C. 258).

morality rather than law; for the concept of murder in ethics must surely involve some element of hatred or malice[8] or, at least, of taking life wantonly or recklessly. But this does not mean that either suicide or "mercy killing" are not, normally, morally wrong — both in the eyes of God and of the society in which we live. The Christian believes that life comes to us as a gift from God; that it is to God and for God that we both live and die; and that it is wrong for man to usurp the divine prerogative by cutting life short in the absence of some overriding justification.

One fallacy that needs to be exploded in this context is the impression that the dying — and, in particular, those dying of cancer — usually suffer from unbearable, or even severe, pain. This is not true. More often than not, the process of dying represents a progressive losing of life's moorings; and even among cancer patients, as many as half may have little or no pain. Experience in St Christopher's Hospice and elsewhere proves that when patients are treated with love and understanding, as persons rather than cases, and when doctors and nurses have time to let them express their anxieties and fears about their families and their future, it is commonly possible to relieve their physical pain with relatively small doses of analgesics — which can be varied by a skilful doctor to meet individual and changing needs. Should this not suffice, moreover, there are a variety of more powerful drugs which can and should be used. This means that all may be enabled to die in peace — although it may take a little time, of course, to get things under control. It is true that few doctors have been adequately trained for such work, and that this quality of care is costly in both material and human resources; but the provision of such training and resources should be given a very high priority in any compassionate society.

There remain, however, a small minority of exceptional cases. Examples commonly given are those of a soldier trapped in a burning tank or a motorist who cannot be extricated from a burning car — where few, if any, would condemn a fellow soldier who shot him dead or anyone else who took appropriate action; and the same principle may occasionally apply in the case of someone dying in great pain when the necessary drugs

[8] Cf. 1 John 3:15.

are simply not available. The problem of what action is right in cases like this is somewhat analagous to what we have already discussed in regard to the birth of a "monster" or hideously misshapen baby — and must, in fact, be left to the discretion and conscience of those concerned. It would be virtually impossible — and most undesirable — to attempt to draft legislation to cover such extreme and exceptional circumstances. Should legal action sometimes be taken against those who act in good conscience in such cases, it seems infinitely preferable to rely on the intelligence of judges and juries, and the flexibility of the law, than to attempt a feat of legal draftsmanship which might open the door to serious abuse.

Two further points of great importance remain for discussion in this general survey of the prolongation and termination of life: the problem of what, precisely, constitutes death (and what should, therefore, be regarded as the purely "artificial" prolongation of life), and a number of pressing problems in connection with transplant surgery. First, then, the question of how death is to be defined today. In the past the arrest of respiration and circulation, or the cessation of breath and heart-beat, have been regarded as the essential, and adequate, criteria; although it has for long been known, of course, that it is sometimes possible for one who is apparently dead to be revived by the "kiss of life" or by heart massage before the cells of his brain have been irreparably damaged by the failure of his heart and lungs. Today, however, it is possible to maintain biological life almost indefinitely by mechanical means, such as an iron lung or a heart machine. But cases have occurred in which these vital processes of respiration and circulation have been kept going artificially for several days, only for an autopsy to reveal that the brain had for long been clearly dead. As a result doctors have commonly come to regard the basic criteria of life or death as neurological: the irreparable breakdown of the central nervous system. On this basis, proof of the permanent cessation of all brain activity tested by a number of readings on an electro-encephalogram would constitute proof of death, and the machines which maintained a purely biological life would then be switched off. But it is now known that breathing can continue *spontaneously* for long after the central nervous system has decomposed, or been destroyed, beyond any known possibility of recovery; and it is significant that Dr Marius Barnard has declared, in the context of his brother Dr Christian Barnard's heart transplantations: "I know in some places they consider the patient dead when the electro-encephalogram shows no more brain function. We are on the conservative side, and consider a patient dead when the heart is no longer working, the lungs are no longer working, and there are no longer any complexes on the E.E.G."[9]

These, then, would appear to be the safest and most comprehensive criteria of death. But it seems that, whereas in Britain death is certified if the spontaneous circulation of the blood and breathing have come to an end and cannot be restored, in most European countries and in the United States death is held to have taken place when brain activity stops.[10] In any case, as Dunstan insists, "the notion of patients continuing endlessly wired up to machines is fictitious", for such machines are — or certainly should be — employed only "as temporary aids, either while remedial attempts are made, or until it is discovered whether the organism, unsupported, is capable of spontaneous function or not. When hope of recovery is gone, the machine is switched off, the artificial support is removed."[11] The most difficult problem is provided by the comatose or decerebrate person, whose "spontaneous vital functions continue, because the lower centres of the brain which control them are the only part of it unharmed; the cerebral cortex, which enables rational self-consciousness, is destroyed". In France and Sweden, it seems, this situation is regarded as actually constituting death. But of such a person Dunstan confidently asserts: "He is not 'kept alive'; he lives. And while he lives, he is entitled to elementary human care, that is nursing care, simply to be fed, turned, kept clean. Whether an attack of pneumonia, say, should be countered with active therapy, like an antibiotic, is another question: appropriate management might rather be to let the body fight its own battle, win or lose — there is no obligation, in such a case, to

[9] Cf. D. M. Jackson, op. cit., pp. 15–17. See also *Decisions about Life and Death* (Church Information Office, London, 1965), pp. 30 f.

[10] *Our Future Inheritance*, p. 92.

[11] Dunstan, op. cit., p. 89.

administer a particular remedy."[12] And he then asks "who are the others said to be 'officiously kept alive?' They are the old, the senile, the mentally handicapped of every age" — and he rightly asserts that "All they claim from us, and about all they are given, is the basic human claim, to shelter, warmth, food, elementary medical and nursing care, and a little companionship." For myself, I am doubtful if *human* life, in any meaningful sense, can be attributed to one whose cerebral cortex has been completely destroyed; but, for the rest, I would fully concur with Dunstan.

But in this context we must, I think, briefly revert to the subject of euthanasia. At present, as we have seen, the debate is concentrated on voluntary euthanasia; but there can be little doubt that, if this were ever to be made legal, there would soon be a demand for further concessions. It would not be long before the argument would be heard that paralysed, incontinent and semi-comatose elderly persons would certainly sign the suitable form if only they were to have a sufficiently lucid interval; so why should not their relatives do for them what they would wish to do for themselves? "That agreed" — to quote a recent article by Gardner[13] — "within a month someone else would say, 'But to expect relatives to make this decision is to impose an impossible emotional burden; let us authorise an official to do this without distressing them.' Naturally parallel arguments would be advanced for the congenitally damaged neonates. It would then be suggested that the problem of approval for the euthanasia of the conscious but incapacitated aged would be even more distressing, and therefore it would be vital to relieve relatives of any involvement in this and have it arranged by some distant office."

I have kept the subject of transplant surgery till the end of this lecture for two reasons. First, because of the fact that, while it is primarily, of course, concerned with the prolongation of life, it is also intimately linked with the termination of life, since most of the organs which are transplanted are taken from dead bodies; and, secondly, because the subject bristles with problems — medical, ethical, social and economic — most

[12] Op. cit., pp. 89 f.

[13] "A new ethical approach to abortion and its implications for the euthanasia dispute", in *Journal of Medical Ethics*, 1975, p. 129.

of which are inherent, to some degree, in many of the other points to which reference has already been made. But while tissue grafting or transplanting is, in a very real sense, a development of contemporary medicine, it is interesting to note that records are apparently available of grafts being carried out as long ago as 3,500 B.C. in ancient Egypt, and of the bitter attacks made by theologians of the sixteenth century on a professor of anatomy at the University of Bologna, Gaspari Tagliacozzi by name, whose techniques have been described in some detail. Tagliacozzi, indeed, provides yet another unhappy example of the occasions on which theologians have gratuitously accused scientists or doctors of impious interference with the handiwork of God and even attributed their work to the inspiration, or intervention, of the devil. It seems, indeed, that between the sixteenth and twentieth centuries tissue grafting was always an emotive issue, and that towards the end of the eighteenth century, for example, the University of Paris banned all such operations.[14] So it was only during the 1950s that corneal grafts received a spate of publicity, followed by an ever-increasing interest in kidney transplantation and then, more recently, in heart transplants.

Corneal grafting or transplantation (used chiefly as a cure for blindness caused by the scarring of the cornea, but also to relieve intractable ulcers) has a longer history of success than any other kind of transplant. The first successful corneal graft between humans was accomplished as long ago as 1905; but it was not until 1937 that eyes obtained from dead bodies began to be used for this purpose. Today more than 1,200 corneal grafts are carried out every year in Britain — although even this falls short by about a half of the number which are needed.[15] But the considerably more difficult technique of kidney transplantation is still more central to our subject, since kidney disease is a relatively common cause of death among young people. In Britain, for example, between 1,500 and 2,000 people between the ages of five and fifty-five die from malfunction of the kidneys every year.

Now there are two different ways of treating a patient suffering from kidney disease: he can be put on a kidney

[14] Cf., for this whole paragraph, *Our Future Inheritance*, p. 83.
[15] Ibid., p. 93.

machine, or he can be given a new kidney from a relative or an unrelated donor, usually a dead donor. A kidney machine mimics the action of a real kidney and filters the blood to remove waste products: but in this treatment by "dialysis" the patient must be connected with the machine for two or three sessions of twelve to fourteen hours a week; connections must be made each time between the blood vessels and the artificial kidney; and the patient must restrict the amount of fluid he drinks and the salt he eats. Inevitably, therefore, his life is considerably restricted in a number of different ways, although the long-term prognosis for patients is good, provided they have adequate supervision. But with up to 2,000 cases requiring treatment every year it would not take long for this to become a huge drain on both medical and nursing resources, and the financial burden on the National Health Service would be crushing. A successful transplant, on the other hand, enables the recipient to live a normal life; and the human and financial resources required may be virtually limited to the initial operation, the cost of which seems to be roughly equal to that of one year's dialysis.

The major medical problems inherent in kidney transplants are how to avoid the rejection of the implant by the body of the donee; how to obtain a sufficient number of suitable kidneys; and how to allocate the available kidneys among the much larger number of those who need them. The first of these problems is inherent in the fact that the system of immunity which guards the body against infection is brought into operation against any foreign body, including the transplant kidney. There are two ways, however, in which this can be minimised: by an attempt "to match as many of the parameters of the immune system of the donor and recipient as is possible before the graft takes place,"[16] and to inhibit the immune system of the recipient by drugs. The difficulty in the first case is, of course, the availability of suitable kidneys, while the disadvantage in the second is that the drugs used to suppress the body's natural inclination to reject the grafted tissue inevitably lower its resistance to all kinds of infection. So the reason why kidney transplantation has been more successful than any other, with

[16] Op. cit., p. 84. I have drawn very extensively on chapter 5 of this book throughout this section of my subject.

the exception only of corneal transplants, is because of the existence of dialysis as an alternative method of treatment both before transplantation and in the event of the transplant being rejected.

The second problem concerns the distressing shortage in the number of kidneys available. With the exception of the small number of living donors — mostly close relatives — who are willing to part with a kidney to save another person's life,[17] the supply is limited to kidneys obtained from the very newly dead. Kidneys for transplantation must be removed within an hour of death, and they can then — at present — be stored for no more than twelve hours on ice, or for twenty-four hours on a perfusion machine, before being transplanted. But these factors inevitably give rise to a number of other problems. First, there is the problem of ensuring that the prospective donor is truly "dead" — in the terms already discussed — before his kidneys are removed; and the major safeguard in this respect is the principle that those who certify his death must be wholly unrelated, professionally, to those who will perform the transplant. Secondly, it is possible, once it has been established that the prospective donor is indubitably "dead" in the sense that his central nervous system has broken down and that his lungs and heart could never again function spontaneously, to preserve the state of his kidneys by keeping him on machines which maintain these functions artificially. While, however, there would seem to be no possible moral objection to this, the necessary liaison between the doctors of the donor and the donee, and the switching off of the machines controlled by the donor's doctors at exactly the time when those of the donee are ready to operate, could clearly, on occasion, give rise to misunderstanding and concern. Yet another problem is the fact that the law in this country decrees that permission to remove organs from a dead body must be obtained either from the donor during his lifetime or from his relatives after his death[18] — and it is exceedingly difficult to do the latter in the minimum time available. Much thought has

[17] In the case of living donors, moreover, there is both a legal and an ethical problem — of which the second is the more serious; for many doctors would refuse to take a kidney from a living donor, except perhaps a very close relative, in view of the danger of rejection.

[18] Cf. the Human Tissue Act, 1961.

been given to this problem, notably by a group set up under the chairmanship of Sir Hector MacLennan in 1969 to advise Parliament on possible ways of amending the Human Tissue Act; but the law has not yet been changed. There are some who feel that it would suffice if the present law were interpreted in a way more favourable to transplantation: by regarding the hospital or its officers at the time of death as "the person legally in possession" of the body, for example;[19] by taking the "surviving relatives" to mean only close relatives; and by interpreting such "reasonable inquiry as may be practicable" in terms of the limited time available. More radically, there is a controversy between those who advocate a system of "opting-in" and those who favour one of "opting-out"; that is, the principle of asking those who are prepared for their organs to be used for transplants after their death to register their willingness, and even to carry a card to this effect, on the one hand, and that of regarding all corpses as available for such purposes except only in the case of those who have specifically registered an objection, on the other. Where the deceased has explicitly consented to the removal of his organs, the scruples of relatives, while deserving of respect, should not be regarded as decisive; and in all cases it could, I think, be plausibly argued that the preservation of life should take precedence over matters of sentiment. There can be no doubt, however, that some of the articles which have been published in the press on this matter have only tended to increase popular hesitation and prejudice; and an attempt to discuss the subject in a way which generates light rather than heat is badly needed.

The third, and potentially the most difficult, problem is how the limited number of available kidneys are to be distributed between those who need them. The obvious answer is on the basis of a clinical assessment of those who are in the greatest need and would be most likely to profit from them. But this, in its turn, largely depends on the opinions of individual doctors. There is also — in regard to availability — the fortuitous factor of where the prospective recipients happen to live. An even more difficult facet of this problem is the degree, if any, to which the value to the community of the individual con-

[19] It seems, indeed, that administrative instructions have recently been given to this effect.

cerned, in terms of his work and ability, should be taken into account. And while the major merit of the National Health Service is that it has greatly decreased the scandal of one treatment for the rich and another for the poor, the fact remains that, unless private practice is to be completely prohibited (and there are a number of serious objections to a State monopoly in medicine), there is always the possibility that a doctor may be able to jump the queue on behalf of a patient who is able to pay for it.

In regard to heart transplants the problems are still more acute. To begin with, heart transplants have been much less successful, hitherto, than kidney transplantations, although the percentage of success, and the likely period of survival, seem to be improving. Again, the medical personnel involved in such an operation is far more numerous; and there can be no question that our very limited resources in manpower and hospital facilities could be much more advantageously used in other ways. It might well be right, therefore, to call a halt, for the present, to all such operations — except for the minimum number necessary to maintain medical research. But this is only an acute example of a vast ethical problem which permeates many of these "issues of life and death" which we have been considering in these lectures — and is inherent, indeed, in almost every sphere of life: the problem of what can be done to try to ensure that all those created in the image and likeness of God are given just and fair treatment. To put the blame for the defects in our society on capitalism or any other political or economic system, for example, is far too simplistic, for inequalities and injustices abound under every form of government. The Christian, in particular, should have no difficulty in locating the root trouble in man's fallen nature and inherent selfishness. And for this there is only one basic remedy: a radical conversion followed by a steady growth in spiritual grace and moral sensitivity. But that does not in any way absolve us from doing all we can, here and now, to promote social justice — or, indeed, from confessing how sadly we ourselves still fall short.

5 Capital Punishment, Violence, Revolution and War

IN THIS FINAL lecture we turn to quite a different aspect of our overall subject "Issues of Life and Death". We shall not be concerned, in this lecture, with complex scientific and medical problems and with the exceedingly difficult moral questions to which our continual progress in research, knowledge and skill — whether as present fact or future possibility — give rise. Instead, we must grapple with the age-old problem of the circumstances (if any) in which it is justifiable deliberately to kill another human being, or to put a human life in jeopardy, not for the real or supposed benefit of the person concerned, but for the alleged good of society as a whole or of some of its members. In the traditional view these circumstances include the legal infliction of capital punishment, the taking of life in legitimate self-defence or in the course of a "Just War", and — somewhat more questionably — in what may be termed a "Just Revolution". But in regard to each of these there is a clash of opinion between those who confidently affirm the right to act in these ways, those who virtually deny it, and those who take an intermediate view.

Let us start with capital punishment, on which both Christian and non-Christian opinion is sharply divided. Most people in this country, I believe, consider it is perfectly right and proper for the State to inflict the death penalty in suitable circumstances, and some Christians even affirm that this is actually mandatory in the case of murder; but there are many others who are passionately opposed to the death penalty ever being exacted, and some who regard it as fundamentally contrary to Christian principles. Those who oppose capital punishment normally do so on a variety of different grounds, although they differ among themselves in regard to what these grounds are and the order of priority in which they place them. Many people would put their primary emphasis on the essential fallibility of human judgment, however many safeguards there may be. The British, for example, commonly pride themselves on the way in which criminal trials are conducted in this country, but there can be no manner of doubt that mistakes are sometimes made — especially in cases in which conviction largely depends on a witness's detailed recollection of exactly what happened or his identification of someone whom he saw only briefly or inadequately. Where this mistake has resulted in a prison sentence, however, the possibility always exists of the sentence being reviewed and the victim of injustice being set free and given some form of compensation; whereas capital punishment is necessarily final and irreversible. But they also commonly insist that the infliction of the death penalty has a coarsening effect not only on the executioner and others immediately concerned but on society as a whole, and that it tends to cheapen human life rather than enhance its value. It would only be justifiable, therefore, if it could be conclusively shown that it was a unique deterrent which must be retained in order to protect society; but no conclusive proof of this can, in fact, be produced. Some, moreover, would add that the attempts which are rightly made to ensure justice — in what may well be a long investigation, a protracted trial, a possible appeal, and a period during which any possible reasons for clemency are considered — are, in a sense, counter-productive in terms of humanity, for there is something almost obscene in carefully guarding a condemned man from disease or suicide simply in order to preserve his life for the hangman's rope. Many today, moreover, insist that the infliction of any form of criminal sanction is justifiable only as an attempt either to reform the criminal or to deter others from committing the same crime, and should never include any element of retribution; and they discount both the moral reformation in the condemned man to which a capital sentence may sometimes contribute and the claim that it deters others any more effectively than would some other form of punishment.

Those who favour capital punishment, on the other hand, commonly argue that, whatever statistics may appear to show, it stands to reason that the death penalty constitutes a unique deterrent and is therefore necessary for the protection of society. This is particularly true, they insist, in the case of a man already sentenced to a long term of imprisonment or setting out to commit a crime which would incur such a sentence — where to shoot his way out, for example, would facilitate escape and could only marginally affect his time in gaol. But they would also argue that a long prison sentence, far from reforming a criminal, is apt either to harden him or result in a basic breakdown of his personality. They would, moreover, maintain — rightly, in my view — that to attempt to exclude any idea of retribution from criminal sanctions is to deprive them of any moral basis and to deny society the right to show how deeply it disapproves of the crimes for which they were imposed or make any attempt to reflect the judgment — and justice — of God; and they would then proceed to argue that only the death penalty can adequately express society's reaction against the deliberate taking of an innocent life, or represent a suitable retribution for callous murder or the more extreme forms of treason.

While, moreover, Christians of a "liberal" persuasion are often in the vanguard of those who regard capital punishment as intrinsically wrong, a biblically-minded Christian can scarcely maintain that it is never legitimate for the State to impose such a sentence. There are clear references in the New Testament to the fact that a ruler or government has a divinely imposed responsibility for the maintenance of justice, the encouragement of virtue and the punishment of vice; and the reference in Romans 13:4 to the fact that a ruler "does not bear the sword in vain" can scarcely be interpreted in terms which exclude his right to inflict the death penalty where this seems to be necessary. For those of us who believe that the Mosaic law was promulgated with divine approval, moreover, the argument that the sanctity of human life necessarily rules out capital punishment is clearly untenable, for that law provides for the infliction of the death penalty for a number of offences. There is all the difference in the world, however, between regarding capital punishment as *permissible* in appropriate circumstances, on the one hand, and *mandatory*, on the other. Few Christians, if any, would consider it right for a modern State to follow the Mosaic law in imposing the death penalty for adultery, blasphemy or disrespect to parents, for example; but there are quite a number who believe that in the case of deliberate murder capital punishment is not only right but mandatory on the basis of Genesis 9:1–7, which they would regard as almost tantamount to a "creation ordinance" rather than a foretaste, as it were, of the Mosaic law. So we must, I think, examine this passage in some detail.

Let me say at once that in my own view the cardinal importance of this passage is the decisive emphasis it puts on the unique value, and consequent "inviolability", of human life. Where, moreover, an innocent human life is in fact deliberately violated, it makes it clear that capital punishment is perfectly permissible in suitable circumstances. But I cannot myself regard this passage as making the death penalty mandatory for murder, for a number of different reasons. First, if capital punishment for homicide should be regarded as a form of "creation ordinance", then why was it explicitly precluded in the case of Cain, whose killing of his brother Abel seems to have been a clear case of murder? Secondly, no distinction whatever is made in this passage between murder and any other form of homicide, for the phrase "of every man's brother I will require the life of man. Whoever sheds the blood of man, by man (or possibly, 'for that man') shall his blood be shed" is all-embracing in its scope. But this manifestly stands in need of very considerable qualification in view of the provision later made for "cities of refuge" for those who killed someone by accident — and, indeed, the limitation of talion to cases of deliberate homicide or wounding in certain other codes of Semitic law. Even the most fervent advocates of capital punishment today, moreover, would reserve it for those whose action was either deliberate or reckless; and many would wish to introduce further qualifications, so that it would be confined to cases which were particularly heinous. One of the major objections to this, however, is the extreme difficulty of drafting suitable legislation. Thirdly, we must take this phrase in its context — which, while it certainly authorises the use of animals, birds, fishes or reptiles for food, explicitly forbids eating any such creature "with its life, that is, its blood"; and which also, by implication, requires a judicial "reckoning" of man's blood at the hand of any beast who has killed him. The prohibition of eating blood is, I think, primarily designed to emphasise the fact that the life of every creature belongs to God, that it should not be wantonly slaughtered, and that its life blood was of symbolic significance in the sacrificial system, while the slaughter of a beast who has killed a man is a common feature in early Semitic law, sanctioned by the Mosaic legislation. As Kidner justly remarks, the purpose of this passage is primarily didactic. "If all life is God's, human life is supremely so", and the principles enshrined in these verses have an abiding validity, however much the means of inculcating them may change; for "one cannot simply transfer verse 6 to the statute book unless one is prepared to include verses 4 and 5(a) with it".[1] Fourthly, it is difficult, if not impossible, not to regard this passage as part of the *lex talionis*, or law of retaliation, common to most primitive law and, especially, to all early Semitic codes — and authorised as such, with certain refinements, in the Mosaic legislation; and few, if any, would regard the *lex talionis* as incumbent on us today. It is important to remember, moreover, that the essence of the *lex talionis* was that the punishment must never exceed the offence — a shocking example of which is provided by the incident recorded in Genesis 4:23, where Lamech boasts that he has killed "a mere lad (Hebrew *yeled*) for a mere wound".[2]

It is, moreover, only the moral law of the Old Testament which is now incumbent on Christians, not the ceremonial law, which found its complete fulfilment in Christ, or the civil and criminal law, in regard to which a Christian is now subject to the laws of the State in which he lives — although it is, of course, his duty to do his best to ensure that those laws are right and just, and even to disobey a law which is flagrantly against his conscience. All in all, therefore, my own view is that

[1] *Genesis: an introduction and commentary* (Tyndale Press, London, 1967), p. 101.
[2] Op. cit., p. 78.

it was right, in all the circumstances, that the death penalty for murder should have been abolished in this country, and that it should not be brought back unless this seems to be the *only* way in which murderous attacks on the police or prison warders, or wanton killing in the course of terrorism, can be restrained. It is true that the purpose of such a limited application of the death penalty would be largely deterrent; but the relevant legislation would be comparatively easy to draft, it could include an express provision which would give a discretion to the court where there was a special reason for clemency, and an element of just retribution would certainly not be lacking.

When we turn to questions of violence and war, it is eminently understandable that the early church should have been strongly pacifist in its attitude and teaching, and equally understandable that the horrors of modern warfare should have inspired a similar sentiment even in non-Christian circles today. All the same, it seems to me exceedingly difficult consistently to maintain, without any qualifications, the doctrine of complete pacifism. It was, indeed, an appreciation of the value and inviolability of an innocent human life, together with a recognition of the strength of the instinct of self-preservation inherent in mankind, that gave rise to the classical doctrine of the right of self-defence — even to the extent of killing an assailant if no other means would avail. And while it may be argued that the dominical injunctions that we should be ready to "turn the other cheek" and not to "resist one who is evil" probably mean that a Christian should always choose himself to suffer death rather than inflict it on someone else, the defence of another person who is being terrorised by a brutal assailant is a very different matter. Few Christians would feel it was right to stand idly by while a woman, a child, or any defenceless person was being killed or tortured, and it would certainly not always be possible to intervene effectively without at least the possibility that the assailant would be killed. If this is true of the private citizen, moreover, then it is even more obviously the duty of the State and the police to protect the innocent, whatever action this may involve. This means that the police — or, indeed, the armed forces — may be compelled to use force, or even to kill, in order to restrain criminals or rioters who cannot be restrained in any other way. And while many who would call themselves pacifists would make a sharp distinction between the use of force to maintain law and order within the body politic, on the one hand, and international war, on the other, the distinction is exceedingly difficult always to sustain. What if the only way to prevent war seems to be the stationing of a "peace-keeping force" between two mutually hostile states? Is it really feasible, in such circumstances, to maintain that the peace-keeping force should carry arms only as a gesture, and should never use them in case of need? Supposing, again, that a pogrom is taking place on the other side of an international boundary, must the police or the army of a neighbouring power always stand idly by, and never intervene to save innocent lives?

The fact, as it seems to me, is that the use of force, particularly when this involves any loss of life, is always evil *per se*, but that there are circumstances in which it may be the lesser of two evils. It is on such a foundation that the doctrine of the "Just War" ultimately rests. For our purpose, the basic criteria of this concept may, perhaps, be summarised as follows: the war must be for a just cause; it must represent the only possible way to restore justice or prevent the continuance of injustice; it must normally (although not invariably) be defensive in character; it must be conducted with no more violence or bloodshed than is absolutely necessary; and it must be waged in the reasonable expectation that the good which it will effect is greater than the evils — human, material and moral — that it must inevitably entail. It is perfectly true that some of these criteria seem singularly remote and idealistic in terms of modern warfare, but what alternative is there? If we accept that, in some circumstances at least, armed resistance to oppression, exploitation or genocide is a lesser evil than submission to tyranny, gross injustice or cold blooded murder, then surely an attempt must be made to apply some such criteria as these. It is, of course, eminently desirable that atomic weapons, poison gas or bacterial warfare should be wholly prohibited; but it is at least arguable that ultimately the only way to prevent the use of atomic weapons is to be in a position in which retaliation would be possible. And while the maxim that no more violence should be used than is absolutely necessary can have little or no tactical relevance to many forms of modern warfare, in which any distinction between combatants and non-combatants becomes increasingly difficult, it might still, I suppose, have some relevance to the *strategic* decision as to whether such methods of warfare are, in the circumstances, absolutely necessary. As I wrote some time ago:

> unless the danger of a possible escalation of a local or conventional war into an atomic holocaust is to be regarded as a conclusive proof that *any* other horror, suffering or injustice must today be accepted as demonstrably preferable to incurring such a risk, then we must cling to the doctrine of the Just War, in so far as this still remains applicable, as infinitely better than a facile surrender to wholly unlimited and unprincipled butchery. The extent to which those who state that the doctrine is now obsolete and irrelevant continue, in practice, still to invoke many of its maxims, seems to me significant.[3]

But many Christians who would regretfully agree that international warfare, however appalling its real and potential evils, might in some cases be preferable to allowing unbridled tyranny to prevail, would feel compelled in conscience to take a different view of civil war or of any violent revolt designed to overthrow an evil government. Nor would their hesitation in this matter be based primarily on the fact that fratricidal strife is usually even more detestable and bitter in its nature and effects than a war against an external enemy, but specifically on what they believe to be biblical principles. And they point, in this context, not only to the example of Jesus himself and his apostles, but to the direct teaching of passages such as Romans 13:1–7 and 1 Peter 2:13–17.

Jesus himself, they insist, took no active steps to abolish such an evil as slavery, or to resist the oppression of an occupying power. On the contrary, he rejected the way of the Zealots and deliberately chose the way of the cross. And his apostles seem to have followed the same path and to have limited their defiance of authority to those circumstances in which it came

[3] *Morality, Law and Grace* (Tyndale Press, London, 1972), p. 90.

into direct conflict with what they regarded as unequivocal divine commands. But surely it is obvious that Jesus himself had a wholly unique task to fulfil? He had not come to deal with questions of politics or social justice, however pressing, but to reconcile sinful men with a holy God — and to do this not, primarily, by proclaiming the love of God through teaching and example, but by bearing their sins "in his own body on the tree". For him to have deviated from this purpose, for whatever cause, would have been unthinkable. And while the apostles could have no part in this unique work of redemption, they too would not let anything deflect them from the all-important commission he had given them to make this redemption known. Deliberately to have challenged the might and provoked the hostility of Rome would have frustrated their very *raison d'être*.

What, then, of Romans 13:1–17? Both passages have been taken by some to teach that every government, whatever its nature, is divinely "instituted" in such a way that the "powers that be" (or the "existing authorities") are *always* entitled to claim a Christian's unswerving allegiance. To rebel against any government is, therefore, to resist "what God has appointed" (or a "divine institution"). But it seems to me that the apparently unqualified injunctions of the first two verses of Romans 13 must be read in the light of the next two verses, which state that those who "continue to do right" need have no fear whatever of those in authority, since they are "God's agents working for your good". It is only wrongdoers who need feel any apprehension, for "government, a terror to crime, has no terrors for good behaviour". Similarly, in 1 Peter 2, while verse 19 certainly teaches that it is praiseworthy for an individual Christian "to endure the pain of undeserved suffering because God is in his thoughts", verse 14 explicitly describes a ruler or governor as God's "deputy for the punishment of criminals and the commendation of those who do right". In other words, the God-given function of secular authority is to discourage social evil and promote virtue and social justice. If, therefore, this situation is turned upside down, and a government becomes so degenerate that vicious laws are promulgated and enforced, corrupt officials are put in authority, injustice is perpetrated and the innocent are persecuted, then it is distinctly arguable that the duty to obey such a government no longer applies — at least in its entirety.

So can there not, in extreme circumstances, be such a phenomenon as a "Just Revolution", on a par with the concept of a Just War? The Old Testament would certainly seem to support this, for there God is consistently portrayed as the Lord of history, who acts both in mercy and in judgment — not infrequently by the hands of men. The revolution of Jeroboam against King Rehoboam, for example, is depicted as an act of divine judgment on King Solomon[4] in which Jeroboam was virtually commanded to rebel against his son,[5] but which, in the event, was precipitated by Rehoboam's own tyranny and folly.[6] Much the same could be said about Jehu's revolution against the house of King Ahab and against Queen Jezebel.[7] There is also the story of Ehud the Benjamite[8] and King Eglon of Moab, of Gideon and the Midianites,[9] and of Daniel's vision in which an angel predicted the fall of King Nebuchadnezzar and proclaimed that this had been decreed "to the end that the living may know that the Most High rules in the Kingdom of men, and gives it to whom he will, and sets over it the lowliest of men".[10]

Some, no doubt, who would have no hesitation about the judgments of God on nations or their governments, even by the hands of men whom he has raised up as his instruments for this purpose, would maintain that a Christian, as such, must have no part in such an act of judgment against his lawful government, however degenerate it may be. But why should God act in this way only through non-Christians? Can we presume to limit God in this way? And how, precisely, is the term "lawful government" to be defined? There are few, if any, governments in the world today which have an impeccable ancestry, for in almost every case the existing government can be traced back to a revolution or *coup d'état*, if one goes back far enough. Sometimes, indeed, this has taken place virtually

[4] 1 Kings 11:11 f. and 2 Chronicles 11:1–4.
[5] 1 Kings 11:29–37.
[6] 1 Kings 12:1–20.
[7] 1 Kings 19:16 and 2 Kings 9:1–37.
[8] Judges 3:12–28.
[9] Judges 6:1–7:25.
[10] Daniel 4:17.

without loss of life; but there can seldom, if ever, be any certainty about this in advance. So it seems to me clear that the possibility of a "Just Revolution" cannot plausibly be denied. The fact that both St Paul and St Peter exhort Christian slaves to obey their human masters even when they are overbearing, and say that those who endure unjust suffering patiently have God's approval, does not necessarily mean that a point may not be reached when an evil and tyrannical government may be toppled by force.

But to say this is a very different thing from supporting what may be called the contemporary "theology of revolution". There are some today who even picture Jesus as a revolutionary,[11] and Colin Morris once wrote: "So the prophetic mantle of Jesus passes to Marx, Lenin and Mao, and then on to Castro, Ho Chi Minh, and Torres. But none can wear it for long. Each has his time of creativity and inspiration, and millions are rescued from oppression and hunger as a result. Such times pass; success destroys the revolution and the struggle must begin again."[12] Thus Joel Carmichael affirms that Jesus' greatest moment of triumph was his "seizure and occupation of the temple in Jerusalem", when he "holds the Roman garrison in check" [*sic*] and is "the head of an organised movement against Rome and against those Jews who are traitors to their country",[13] while Albert Cleage asserts that "Jesus was the coloured leader of a coloured people carrying on a national struggle against a white people . . . The activities of Jesus must be understood from this point of view: a man's effort to lead his people from oppression to freedom."[14] But to write in this way is to substitute fiction for history, political phantasy for scholarly objectivity. Richard Shaull, again, insists that "the God who is tearing down old structures in order to create the conditions for a more human existence is himself in the midst of the struggle", so "the Christian is called to be

[11] Cf. S. G. F. Brandon, *Jesus and the Zealots* (Manchester University Press 1967).

[12] *Unyoung, Uncoloured and Unpoor* (Epworth Press, London, 1969), p. 151.

[13] "L'Epée de Jésus", in *Nouvelle revue française*, 1961 (quoted in English by Jacques Ellul), *Violence: Reflections from a Christian Perspective* (SCM Press, London, 1970).

[14] *Le Monde*, January 1968 (as quoted by Ellul, op. cit., p. 48).

fully involved in the revolution as it develops. It is only at its centre that we can perceive what God is doing" and can "discover that we do not bear witness in revolution by preserving our purity in line with certain moral principles, but rather by freedom to be *for man* at every moment".[15] And Father Maillard even goes so far as to say: "If I noticed that my faith separated me by however little from other men and diminished my revolutionary violence, I would not hesitate to sacrifice my faith."[16]

But a man who can write like this can have no real knowledge of the faith he would so easily abandon. To the Christian his faith is not dispensable, revolution can never be a "theology", and violence — at best — can never be other than the lesser of two evils. To opt for revolution can only be an agonising choice, and the alternative must be grim in the extreme. To protest in a constitutional way against what one believes to be wrong seems to me always legitimate and sometimes a duty. In some circumstances, moreover, to protest by means of a "sit-in"[17] or demonstration may be justified, although care should be taken not to harm, or even seriously inconvenience, innocent persons. But acts of positive violence can, I think, be justified only where injustice has reached the level of a basic denial of human rights and dignity, and where no other means of redress is possible. Even then, there must be a genuine prospect that the revolution will succeed, that a better government can be set up in the place of the one that is to be overthrown, and that the suffering inevitably involved in the process will be less than the suffering it is designed to remedy. Again, the means to be used must not involve more violence than is strictly necessary; and an attack on armed forces or ammunition dumps, for example, is an entirely different thing from planting bombs where they may cause death, mutilation or injury to innocent persons. The time has surely come when acts of indiscriminate terrorism, hi-jacking and kidnapping as forms of political protest and pressure should be treated as international crimes. To react in such a

[15] "Revolutionary Change in Theological Perspective", in *Christian and Social Ethics in a Changing World*, ed. by J. C. Bennett (SCM Press, London, 1966), pp. 33 ff.

[16] Quotations from *Frères du Monde*, taken from Ellul, op. cit., pp. 56 ff.

[17] Acts 16:37 seems to record the first Christian "sit-in"!

way to some real or alleged injustice — or even to some manifest and gross injustice over which the victims of such acts of terrorism have no control — should be condemned as the sheer anarchy that it is. But we must be careful neither to follow Jacques Ellul into the moral confusion of labelling indiscriminately as "violence" both revolutionary terrorism and the legitimate use of force by the State, nor the opposite mistake of failing to sympathise with those who feel driven to react, in a way one cannot altogether condone, to the tyrannical use of what can only be described as governmental violence.

Finally, we must, I think, look back over these lectures as a whole. Time and again we have had to face exceedingly difficult problems — quite literally issues of life and death — in some of which a clear-cut moral judgment is virtually impossible. But surely there is a very real sense in which every advance in scientific and medical knowledge is God-given, not wrested from an unwilling deity; and the Christian believes that it should be used, reverently and responsibly, to further the divine purposes in creation and redemption. The problem, of course, is how we can know what these purposes are, and how they are to be furthered in any concrete case. In the past, reference has continually been made in this context not only to the Bible but to Natural Law; that is, the belief that God has written his laws, as it were, into the very structure of the universe, and that man, as a rational being, can — in part — "read them off".

But part of our present bewilderment comes from the fact that man is now able to manipulate nature to a degree which was not possible in the past; and while we may justly say that the exploitation of chickens and calves, for example, in a way which deprives them of daylight and freedom of movement, is "contrary to nature", we may well take a different view of the conquest of pain, disablement, frustration and disease by methods which clearly transcend "nature" as we have known it in the past. As I observed in a previous lecture, St Paul makes it very clear that the divine approval originally pronounced on the natural creation as being "very good" no longer holds true, for today it is "subjected to futility" and "groaning in travail".[18] One day, we are told, it will be set gloriously free;

[18] Cf. Romans 8:19–21.

but even now man is becoming progressively more and more competent to do this in a minor way. So, for the Christian, the problem in every case is not only whether this possibility or that will ultimately do more harm than good, but the still more basic question whether it is, or is not, in accordance with the divine will. But hitherto we have given no adequate attention to how he is to find the answer to this question.

It has often been remarked that there are, basically, three sources of authority in religion: revelation, tradition and the reason (or "inner light") of individual men and women. This is certainly true for the Christian, who stands in need of all three. But only confusion would result from three sources of authority which were on a completely equal footing; and Christians, down the ages, have differed considerably in the order of priority which they accord them in practice. Almost all would, in theory, give priority to the Bible, but some would put such an emphasis on their belief that the Bible must be authoritatively interpreted by the Church that they would accord tradition an almost equal status, while others place such reliance on their own minds or on what they believe to be the direct guidance of the Holy Spirit that these, in practice, constitute for them the final arbiter. But the Evangelical, as I understand it, feels that he must always give an absolute priority to the Word of God, both incarnate and written. Even so, that Word has to be interpreted and applied; for God treats us as adults rather than children, and has not given us precise and detailed instructions about everything. Next, then, the Evangelical turns to tradition; for only a fool disregards the accumulated wisdom and experience of the past. In the last resort, however, he believes that the individual must seek the illumination of the Holy Spirit in his own mind and heart; and, although he will hesitate long before he parts company with tradition, yet where he is convinced that this is wrong he must follow his own conscience as to what the Bible teaches, or what he believes to be consonant with its teaching.

But how does all this apply to the problems we have been discussing in these lectures? In these matters, too, we must, as I see it, wholeheartedly accept the biblical revelation as authoritative — from Genesis to Revelation. But the trouble is not only that we often do not know precisely what some verse or passage means, and how it is to be applied, but that many of our contemporary problems were unknown to the biblical writers — and so are covered, at best, by principles which are laid down in the Bible rather than by its detailed precepts. And this, of course, applies still more to tradition, to which I would not, in any case, accord the same category of inspiration and authority. Past generations lived in a very different world; even General Councils, we know, have erred; and the traditional teaching of the Church has been somewhat obsessed, in these questions, by the concept of Natural Law in its objective sense of principles written into the very structure of nature as we know it — and nature, as we have seen, can now be manipulated, to a considerable degree, by man, whether for good or ill. But the other sense of Natural law — that man is a rational being who can in some degree perceive the will of God, and on whose heart the requirements of that will are in some part written[19] — is as true today as it ever was. And this is applicable not only to the Christian, but to man as man. What is essential, as I see it, is that we should remain humble, reverent and teachable — for if General Councils have erred, then we ourselves are manifestly liable to make mistakes. That is why I have very seldom, in these lectures, made any attempt to give dogmatic answers, but have contented myself with outlining one problem after another and then making very tentative suggestions as to the right solution.

But is there no more that can be said? Is there nothing that the Christian, as such, can do except hope that he will be rightly guided? — and here I think particularly of the doctor or scientist who has to make practical decisions, as well as the community whose common mind should guide him and inform the law. About this I think a little more can, indeed, be said. We are concerned with two closely related problems: first, how to distinguish between what is right and what is wrong; and secondly, in more general terms, how to know the will of God. Now in regard to the first of these problems it seems to me most significant that we are told, in Hebrews 5:14, that one of the distinctive marks of mature Christians is that they "have their faculties trained by practice to distinguish good from evil", and that the previous verse makes it clear that the

[19] Cf. Romans 2:15.

pre-requisite for this is a constant feeding on, and resultant skill in understanding and applying, "the word of righteousness". It is a regular and intelligent study of the Bible as a whole, rather than a reliance on some particular verse or verses in isolation, that represents one side of this "training" which enables us to distinguish what is right from what is wrong — or, at times, what is the lesser of two evils in those situations in which there is no "absolute" right (whatever the advocates of "Situation Ethics" may claim) in the circumstances of this very imperfect world. But this training is also, we are told, the result of "practice" — which can, I think, only be acquired by experience in dealing with the facts of daily life. We are concerned with the problems of men and women, so we need to know both the character of God and the nature of man made in his image; and this is where the experience of the Christian doctor, pastor, scientist, lawyer, nurse and social worker can complement — and, at times, even correct — the conclusions that the theologian, philosopher or logician reaches in his study. We must never abandon those "Maker's Directions" which provide us with ethical standards of abiding validity; but man made in the image of God means, in practice, a vast variety of men and women with very different natures and circumstances which must be taken into account in applying these moral "absolutes" to their individual problems.

Then, again, it seems clear that we can only know the will of God if we remain essentially teachable. It is not for nothing that we are told that it is the "meek" that God will "guide in judgment". Even those with a profound knowledge of the Bible can all too easily acquire the rigidity of a closed mind. This is why St Paul tells us in Romans 12:1–3 that in order to "discern the will of God, and to know what is good, acceptable and perfect" we need first to offer our very selves to him as "a living sacrifice", and then to let our minds "be remade" day by day. We must, of course, thankfully accept the expertise and enlightenment for which we stand in constant debt to our non-Christian colleagues, for their knowledge and skills are themselves gifts of God's "common grace" which he bestows on man as man. But, left to ourselves, we should all too easily grow more and more adapted, or "conformed", to the "pattern of this present world", and the only remedy for this is a continual process of what the Apostle terms being "transformed" by means of the "renewal" of our minds. It is essential, then, that our minds should always remain open and receptive to the new light God gives us through his revelation of his own nature and will in the Bible and through our growing knowledge of the nature of man made in his image and likeness.

But we started these lectures by concentrating our attention on the unique status and dignity of man made in the image of God, and it is on this same note that we must bring them to an end; for it is this which must always inform our attitudes and guide our decisions in regard to these vital issues of life and death. After all, the problems with which we have been wrestling find their genesis in the unique value of human — as distinct from animal — life. This certainly does not justify the cruel way in which we often exploit animals for our greed or convenience, or use them for scientific and medical experiments which are not strictly necessary. But it does, I believe, justify the use of animals to test out drugs, for example, provided this is done in a way which causes as little suffering as possible — or for genetic experiments which contribute to vital knowledge rather than idle curiosity, since an unfortunate result can be largely remedied by a painless death. But man is on quite a different footing.

Now the Bible makes it abundantly clear that man as we now know him is a fallen creature, who stands in need, at every point, of the redemption that is in Christ Jesus. Yet sin, as we have seen, has not wholly effaced the image of God in man, and that image *can* be fully restored — partly in this life, and perfectly hereafter. If we wish to see man as God intended him to be, we must look at the human life of Jesus, lived in continual dependence on, obedience to, and communion with, his heavenly Father. And if we want to see man as those "many sons", whom it is God's supreme purpose to "bring to glory", will *all* one day be, then we must look at our Lord Jesus Christ after his resurrection and ascension. Part of our trouble today is that we are obsessed with the importance of this present life and regard its termination as the ultimate tragedy. We are so afraid of the accusation that our religion is only concerned with "pie in the sky when we die", and we are so conscious of our urgent duty to do all we can for our fellow human beings here on earth, that we too often forget the "glory" to which we should look forward — not as a form of escapism, but as an inspiration for action. It was with an eminently practical purpose in view that C. S. Lewis wrote:

> It may be possible for each to think too much of his own potential glory hereafter; it is hardly possible for him to think too often or too deeply about that of his neighbour. The load, or weight, or burden of my neighbour's glory should be laid daily on my back, a load so heavy that only humility can carry it, and the backs of the proud will be broken. It is a serious thing to live in a society of possible gods and goddesses, to remember that the dullest and most uninteresting person you talk to may one day be a creature which, if you saw it now, you would be strongly tempted to worship, or else a horror and a corruption such as you now meet, if at all, only in a nightmare. All day long we are, in some degree, helping each other to one or other of these destinations. It is in the light of these overwhelming possibilities, it is with the awe and the circumspection proper to them, that we should conduct all our dealings with one another, all friendships, all loves, all play, all politics. There are no *ordinary* people.[20]

So surely, in ending these lectures, we can let ourselves go, as it were, for a minute or two. Those whom Christ has redeemed are, in sober truth, on the way to a glory we cannot begin to understand. "The sufferings we now endure," St Paul assures us, "bear no comparison with the splendour, as yet unrevealed, which is in store for us." It is for this, indeed, that the whole of creation is waiting; and we ourselves, who "have the firstfruits of the Spirit", look forward with eager longing to the "redemption of our bodies". Meanwhile "we know that in everything God works for good with those who love him, who are called according to his purpose. For those whom he foreknew he also predestined to be conformed to the image of his Son, in order that he might be first-born among many brethren. And those whom he predestined he also called; and

[20] "Weight of Glory", in *Screwtape proposes a Toast*, (Fontana).

those whom he called he also justified; and those whom he justified he also glorified." And St John strikes precisely the same note when he exclaims: "How great is the love that the Father has shown us. We were called God's children, and such we are . . . Here and now, dear friends, we are God's children; what we shall be has not yet been disclosed, but we know that when he appears we shall be like him, because we shall see him as he is. Everyone who has this hope set on him purifies himself, as Christ is pure."

Author Index

Subject Index